Kirsten Rain

Winning the Hardware–Software Game

Winning the Hardware–Software Game

Using Game Theory to Optimize the Pace of New Technology Adoption

Ruth D. Fisher

PRENTICE
HALL

Upper Saddle River, NJ • Boston • Indianapolis • San Francisco
New York • Toronto • Montreal • London • Munich • Paris • Madrid
Capetown • Sydney • Tokyo • Singapore • Mexico City

The publisher offers excellent discounts on this book when ordered in quantity for bulk purchases or special sales, which may include electronic versions and/or custom covers and content particular to your business, training goals, marketing focus, and branding interests. For more information, please contact:

U.S. Corporate and Government Sales
(800) 382-3419
corpsales@pearsontechgroup.com

For sales outside the United States please contact:

International Sales
international@pearsoned.com

Visit us on the Web: informit.com/ph

Library of Congress Cataloging-in-Publication Data

Fisher, Ruth D.

Winning the hardware–software game : using game theory to optimize the pace of new technology adoption / Ruth D. Fisher.
p. cm.
Includes bibliographical references and index.
ISBN 978-0-13-700282-5 (pbk. : alk. paper)
1. Information technology—Management. 2. Technological innovations. 3. Game theory. I. Title.
HD30.2.F575 2009
338'.064015193—dc22

2008053883

ISBN-13: 978-0-13-700282-5
ISBN-10: 0-13-700282-3

Text printed in the United States on recycled paper at Courier in Stoughton, Massachusetts.
First printing, February 2009

To Dad

Contents

Preface

During the summer of 2005, my friend and intellectual provocateur Lloyd Nirenberg, an expert in wireless technologies, e-mailed me an article describing the difficulty HDTV manufacturers were having getting viewers to adopt HDTVs. I responded by telling him that it was a classic case of the chicken-and-egg problem, and that I had been following the issue for quite some time. He in turn asked me if I could model the game.

In essence, a *chicken-and-egg problem* occurs when system innovators introduce hardware components of new hardware–software systems into the marketplace, but adoption is *held up*, or delayed, by users who refuse to adopt the new system until there is enough content available to go with the hardware, while software providers refuse to supply content until enough users have adopted the new system and can access the content. I refer to the dynamics during the adoption process of new technology systems among hardware suppliers, software suppliers, and users as the *Hardware–Software Game*.

I spent about a month or so playing with equations that captured the dynamics of the Hardware–Software Game, and I discovered some interesting insights, which I shared with Lloyd. He was very excited and told me that I should write a book. My reaction to this suggestion was great skepticism, because I figured that economists and technologists must surely already have studied the issue. I started researching the subject and, much to my surprise, while I found acknowledgment of the chicken-and-egg problem, I could find no systematic analysis of it.

I found many discussions of specific case studies, such as the battle between VHS and Betamax during the 1980s and the progression of the next-generation DVD standards war being waged at the time (2007–2008) between Blu-ray and HD-DVD. There was also plenty of analysis by researchers of other issues tangential to the Hardware–Software Game, such as network effects and product lifecycles. But all I found on the topic of the chicken-and-egg problem itself were brief descriptions of the issue, with only scattered mention of a couple of different potential solutions, which would solve the problem only under certain situations.

After I continued to play with my model, it eventually became clear to me that the holdup problem associated with the successful adoption of new hardware–software systems depends on three primary factors: the existence of networks and network effects, the value proposition offered by new systems, and adoption and validation of new technologies by early adopters.

So then I set about researching these three topics in more detail. I found it interesting that there appeared to be three general approaches to each of these problems: (1) a social or anthropological approach, which deals with the diffusion of information through personal interactions; (2) a business and strategy approach, which analyzes the impacts of advertising, pricing, and validation on the speed of new technology adoption; and (3) an economic approach, which deals with the dynamics and implications associated with specific aspects of technologies, such as network effects, standardization and compatibility, and path dependence.

Even more interesting was the fact that any given piece of research tended to analyze the issues using only one of the three approaches. In other words, I could find little research that combined sociological, strategic, and economic perspectives into a single analysis. This observation became even more remarkable to me after I realized that while each of the three approaches has something unique to offer and explains critical aspects of the Hardware–Software Game, no one discipline alone could fully explain the problem in general (as opposed to being able to explain specific instances of the problem). It was only when I integrated ideas from all three disciplines that I was satisfied that my model was able to simultaneously capture all the major issues. Finally, then, I was able to correctly and systematically examine the dynamics of the general problem, from which I could discern insights and offer potential solutions.

So what this book offers is a systematic analysis of the holdup problem in the adoption of next-generation hardware–software systems. The analysis combines aspects of sociology, business strategy, and economics to examine the underlying dynamics of the adoption process, and it offers methods by which system innovators can provide incentives for content providers and system users to adopt next-generation systems sooner than they might otherwise.

The topics covered in the book apply to an entire genre of technology systems, namely, those that require users to combine hardware and software components to access the technology's consumption experience. As technology systems have evolved and markets have become increasingly global and competitive, these types of systems have become more prevalent, and they will continue to play a large role in technology markets as systems continue to evolve. In particular, the market for consumer electronics intimately and strongly depends on these dynamics. Those who make and sell consumer electronics in general, especially during weak economic times, should be particularly interested in these results and methods. More generally, most communication and transportation systems are hardware–software systems, so participants in either of these industries would benefit from the analysis presented in this book.

Although the book is based on economic and mathematical modeling, it has been specifically written to appeal to an audience without a strong technical background in economics or mathematics. In particular, the book provides numerical examples and uses real-world case studies (for example, adoption of cell phone systems, computers and printers, PDAs, CDs and DVDs, HDTV, and so forth) to illustrate the more theoretical concepts, so that even less technical readers can obtain a good understanding of the book's basic message. At the same time, however, more technical readers have access to the mathematical model, which is presented in Appendix A. Additionally, Appendix B provides readers who are interested in further pursuing its ideas with a list of resources, and Chapter 7 discusses specific tools for applying to their own technology systems the ideas presented throughout the book.

The book is organized as follows. The first two chapters provide readers with a brief discussion of the fundamentals they need to understand the analysis of the Hardware–Software Game. Chapter 1 provides a brief discussion of networks and network effects, switching costs, and technology compatibility and ends with a brief discussion of the role network effects play in the Hardware–Software Game. Chapter 2 provides a brief description of technology adoption lifecycles and ends with a brief discussion of the role technology adoption lifecycles play in the Hardware–Software Game.

The next two chapters provide descriptions of each of the players in the Hardware–Software Game. Chapter 3 builds upon the basics discussed in Chapters 1 and 2 to describe the nature of user demand for system hardware and software. The chapter also provides numerical examples to give readers a more tangible understanding of how the demand functions work. Again building upon the basics from Chapters 1 and 2, Chapter 4 discusses the particular issues that affect the supply of system hardware and software, together with the choices each type of provider must make when new technology systems enter the market.

Chapter 5 presents the main analysis. It describes a simulation model of the Hardware–Software Game, which is then utilized to characterize the game's dynamics.

More specifically, the chapter provides a systematic analysis of the Hardware–Software Game and describes the outcomes for each of the players associated with different scenarios. Chapter 6 uses the analysis of the Hardware–Software Game from Chapter 5 to provide methods that system innovators can use to modify market environments so that software providers and users will be led to adopt new technology systems sooner than they would otherwise.

The last chapter of the book, Chapter 7, provides a summary of the key concepts presented in the book, tools that readers can use to apply the concepts discussed in the book to their own technology systems, and a brief description of further areas of research.

Acknowledgments

Although the cover of the book designates a single author of this work, the book would not have come into existence without the initial suggestion and then subsequent prodding, direction, and continual encouragement by my friend and colleague Lloyd Nirenberg.

Also indispensable were the unequivocal faith and support and persistent nurturing of my mom, Judy Fisher, my sister, Gwen Fisher, and my brother Arthur Fisher, and of my dear friend Tony Bozzuto. I would further like to acknowledge the invaluable advice Gwen provided me about writing style and John Hollis provided me about the general book-writing process. Next, I am extremely appreciative of the comments provided by the reviewers of my book, which helped me make it much better than it would have been otherwise, and I thank Bernard Goodwin at Prentice Hall for his considerable efforts and support in getting my book published.

And finally, I'd like to thank my dad, Norm Fisher, who always has been, and always will be, my number-one fan.

About the Author

Ruth Fisher, Ph.D., is founder and principal of QuantAA, an economic consulting firm that applies advanced quantitative assessment and analysis to help clients make better decisions. She has had a lifelong fascination with the discovery, development, and commercialization of new ideas.

Introduction

Have you ever had a piece of recorded music and no music player on which to play it? There you were, with your favorite Bob Dylan CD, and not a CD player in sight. Or maybe you have an old cell phone sitting in a drawer. It never rings because nobody knows it is there but you.

In both of these examples, you need both hardware and software or content or connection services (jointly referred to as *software*) to make the thing work. With only a CD and no hardware on which to play it, there is no music. With no connection services from a phone company, the phone doesn't work either. In each example, either the hardware (the CD player) or the software (the phone service) is missing.

Different platforms such as the cell phone or disc player use different types of technologies to work. A cell phone needs a service provider, whereas a CD is a media source. In both cases, the hardware won't work without this missing piece, what I call *software*.

As technologies evolve and become ever more complex, these *hardware–software* technology systems are popping up in ever more locations. These systems require consumers to simultaneously use both a hardware platform together with a software or content component to enjoy the *consumption experience* provided by the complete technology system. More examples of technology systems include

- Audio/video systems, including CDs, DVDs, TVs, and iPods
- Gaming systems, including Sony PlayStation; Microsoft Xbox; and Nintendo GameCube, Wii, and Game Boy
- Automobiles, which require fuel and maintenance
- Computer systems
- Communication systems, including facsimile machines, cell phones, and PDAs
- Transportation systems, which require vehicles (planes, trains, and automobiles, for example) and roadways (routes, tracks, and roads)

- Automatic teller machines
- Data collection and storage systems, including radio-frequency identification (RFID) systems, scanners, and smart cards
- The NYSE, NASDAQ, eBay, and any other system that requires system infrastructure and equities or products to trade or sell within them

In most cases of technology systems, users buy hardware from one company and software from other companies. For example, RCA, Sony, and Magnavox sell TVs, and various broadcast, cable, and satellite companies sell programming services. Sony, Panasonic, Emerson, and Philips sell CD players, whereas EMI, Warner Brothers, and Columbia sell music CDs. And Motorola, Qualcomm, and Samsung sell cell phones, and Verizon, Sprint, and T-Mobile sell local and long-distance services.

In these examples, in which separate companies provide the hardware and software components, the marketplace will adopt new or updated system technologies only when the actions of three different groups synchronize: hardware manufacturers; software, service, and media suppliers; and potential adopters. More specifically, *hardware manufacturers* are suppliers of the new system hardware, which includes any system software needed for the basic functioning of the hardware. In contrast, *software providers* are suppliers of new system software, accessories, services, or other content whose products enhance the value of the new system hardware, thereby attracting new users to the system. Both of these groups sell to the *potential* adopters, such as large companies, governmental agencies, or regular people like you and me.

When a new technology system is introduced into the marketplace, these three groups will act independently of one another and do whatever is in their own best interest. Hardware manufacturers will set prices to encourage potential adopters to upgrade to their new systems in a way that will maximize the manufacturers' profits, regardless of what may maximize profits for software providers or optimize value for users. Similarly, software providers will continue to supply content for already existing systems, or they will switch over to providing content for newly introduced technology systems, whichever will earn them the most profits, regardless of what may be best for hardware manufacturers or users. And potential adopters will continue to use installed technology systems, or they will upgrade to newly introduced systems, whichever will give them the most value, regardless of what may generate the most profits for hardware manufacturers or software providers.

The adoption of high-definition television, or HDTV, is a good example. The adoption of HDTV by Americans has been remarkably slow, because the interests of the three sets of players—manufacturers of HDTVs, providers of program broadcasting, and viewers like you and me—have been at odds with one another. The manufacturers of HDTV sets would have liked to sell their TVs to viewers much more quickly than they have been able to, because they would have generated more profits with a speedier pace of adoption. The U.S. government wanted U.S. television viewers to adopt HDTV more quickly than they have, because it wanted to free up the spectrum currently being used for analog television broadcasting for other purposes. In other words, HDTV manufacturers and the U.S. government wanted the adoption curve for HDTV to follow the path of the curve on the left in Figure 0-1. Unfortunately for them, however, the reluctance of U.S. viewers and television broadcasters to upgrade to HDTV means the adoption curve for HDTV in the U.S. market looks more like the curve on the right in Figure 0-1. As it happened, though, it was in the broadcasters' best interest to continue to supply old TV programming instead of upgrading to HDTV. Similarly, TV viewers decided that they would rather continue to watch their current analog (or non-HDTV-compatible) TVs instead of spending money to buy a new HDTV. The conflicting interests of the three groups have caused the new technology of HDTV to be adopted at a snail's pace.

Digital television, of which HDTV is a particular type, was introduced into the previously all-analog U.S. marketplace during the 1990s. "In 1997, Congress set a December 31, 2006, deadline for the transition to all-digital broadcasts."[1] Certainly,

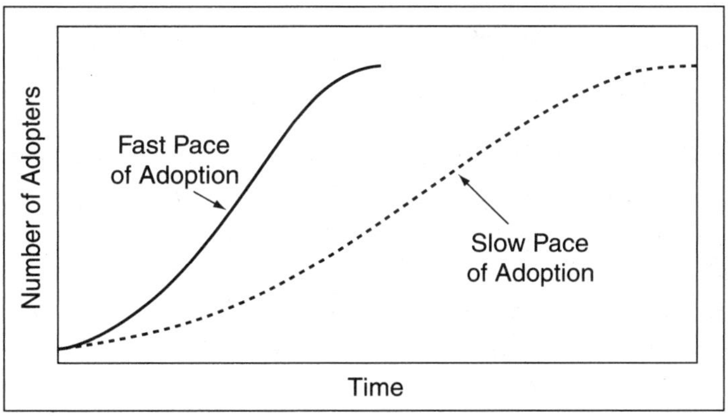

Figure 0-1 Technology adoption curves

1. Schatz 2005.

this deadline whooshed past before everyone adopted HDTV. Consumers have been slow to upgrade their hardware from the traditional analog technology because prices of HDTVs have been too high to justify buying them because of the minimal amount of digital programming available. The hassle and cost of upgrading to HDTV, including getting rid of the old set, outweigh the sparse benefits for most people. For their part, TV stations have been slow to provide digital broadcasts because of the high cost of changing from analog to digital, especially to supply only a few viewers who have HDTVs.

Was there any way HDTV manufacturers could have sped up the pace of adoption of HDTV in the U.S. marketplace? That is precisely the question this book seeks to answer.

0.1 Game Theory and the Chicken-and-Egg Problem

I use game theory to figure out how to speed up the pace of adoption of new system technologies like HDTV. To do this, I need to better understand the incentives each entity (hardware manufacturers, software providers, and users) faces when a new technology system is introduced into the marketplace. This leads me to turn to game theory in search of some answers. *Game theory* examines situations in which entities

1. Act with, and react to, each other, independently from and strategically with one another, doing what is in their own best self-interest
2. Receive payoffs that are dependent upon the actions of the other players

The concepts underlying game theory have been used since time immemorial:

> *The ideas underlying game theory have appeared throughout history, apparent in the bible, the Talmud, the works of Descartes and Sun Tzu, and the writings of Charles Darwin ... While many other contributors hold a place in the history of game theory, it is widely accepted that modern analysis began with John von Neumann and Oskar Morgenstern's book,* Theory of Games and Economic Behavior *[1944] and was given its modern methodological framework by John Nash building on von Neumann and Morgenstern's results.*[2]

Since von Neumann and Morgenstern's work in the 1940s, the game theory framework has become more widely and deeply applied to the point that "game theory is a

2. Shor 2006.

sort of umbrella or 'unified field' theory for the rational side of social science, where 'social' is interpreted broadly, to include human as well as non-human players (computers, animals, plants)."[3] More recently, game theory has been used to understand behavior in biology, economics, politics, computer science, philosophy, sociology, and psychology,[4] to such an extent that

> *since at least the late 1970s it has been possible to say with confidence that game theory is the most important and useful tool in the analyst's kit whenever she confronts situations in which what counts as one agent's best action (for her) depends on expectations about what one or more other agents will do, and what counts as their best actions (for them) similarly depend on expectations about her.*[5]

In the HDTV game, hardware manufacturers act by manufacturing specific numbers of television sets and setting their prices. They react to software providers and viewers by providing fewer sets and/or charging lower prices when there is not a lot of HDTV programming available and/or viewers are not buying HDTV sets. Software providers act by choosing to provide none, some, or a lot of HDTV programming. They react to viewers by increasing the amount of HDTV programming available as more users adopt HDTV or by charging more for HDTV programming. Viewers act by choosing either to stay with their analog systems or to upgrade to HDTV. They react to hardware manufacturers by upgrading faster when the price of HDTV sets is lower, and they react to software providers by upgrading faster when there is more digital programming and when the price of digital programming is lower.

These three sets of players do what is in their own self-interest—that is, what will maximize their own profits or value—regardless of what might be best for either of the other two sets of players. Ultimately, though, the amount of profits HDTV manufacturers and television broadcasters generate and the amount of value viewers receive will depend on the actions the other two sets of players take, that is, the prices hardware manufacturers charge, the amount of programming broadcasters make available, and the timing of users' adoption.

---

3. Aumann 1987.

4. See, for example, Maynard Smith 1982 (biology), von Neumann and Morgenstern 1944 (economics), Downs 1957 (politics), Ben-David et al. 1994 (computer science), Quine 1936 and 1960 (philosophy), Swedberg 2001 (sociology), and Camerer 2003 (psychology).

5. http://plato.stanford.edu/entries/game-theory/.

In essence, then, this problem of trying to coordinate the actions of independent (1) providers of system hardware and (2) providers of system software with (3) potential users is actually a strategic game that plays out when the new Xbox hits the market, Apple introduces a new gadget, or more generally, anytime new system technologies are introduced into the marketplace.

The problem that often appears in these hardware–software games is the classic *chicken-and-egg problem*: Software and media providers are reluctant to supply content early on because too few users have adopted the technology systems to make the new markets profitable for software providers to enter, and users are reluctant to adopt the new systems early on because there is not enough content to make their consumption experiences worthwhile. The chicken-and-egg problem can result in long delays before technology systems are successfully adopted, if adoption is not completely stymied altogether.

For system technologies, the trick, then, is to get all three sets of players to play the game at the same time. The hardware manufacturers must provide the new technology, and *at the same time* the content providers must find it profitable to supply content for the new technology, and *at the same time* users must find it worthwhile to adopt the new technology. Only when all three players are simultaneously on the same page will the technology become adopted at a reasonable pace.

There has been a variety of previous studies of the adoption paths of hardware–software system technologies, but the studies tend to examine adoption paths of specific technology systems on a *case-by-case basis*, providing insights about the particular cases studied. For example, it has been established that new systems that are compatible with existing systems tend to get adopted much more easily, as do new systems that provide a new "killer app." However, there is scant research either on the economic structure of the general dynamics between hardware and software suppliers itself during the early stages of adoption of new technology systems, or on methodical or systematic analyses of how hardware providers might speed up adoption of their new systems under various situations and/or market scenarios.

And that is where this book comes into play. Specifically, this book examines the holdup problem encountered during the adoption process of a particular class of new technology systems. The class consists of systems in which (1) both hardware and software components are required to produce the consumption experience for users of the technology systems, (2) the hardware and software components for the technology systems are supplied by providers who act independently from one another, and (3) the technology systems exhibit network effects. I refer to the interactions of these independent suppliers of hardware and software components during the course of introduction of new technology systems as the *Hardware–Software Game.*

This book proposes to provide a *systematic* analysis of the structure of the Hardware–Software Game. From this comprehensive analysis, I generate insights for different systems and market scenarios. I use a simulation model based on game theory to examine the economic structure of the Hardware–Software Game, to see what is in the best interest of each set of players, to determine if and when they will be led to adopt a new technology system on their own, and to see what the hardware manufacturers can do to get their systems adopted sooner rather than later. Once the structure of the game has been examined, it will be easier to understand such issues as the following:

- Why do some great systems get adopted, whereas others do not?
- Which forces drive adoption by users and content providers?
- How can system innovators speed up the pace of adoption of their new technology systems?

After analyzing the structure of the game and better understanding the underlying dynamics, I am then able to propose means by which new system innovators can help speed up the pace of market adoption of their new systems.

0.2 Organization of the Book

Without a clear and connected understanding of some basic concepts, readers will have a more difficult time grasping the dynamics of the game. As such, the first part of the book provides the building blocks for the analyses presented in the second part of the book; that is, the earlier chapters, Chapters 1 though 4, provide readers with the information they need to know about the basic elements of technology systems, technology adoption, and the players themselves that characterize the venue in which the Hardware–Software Game is played.

Network systems and network effects are central features of the technology platforms within which the Hardware–Software Game is played. Network effects play a crucial role in deterring software providers and users from immediately adopting new technology systems. As such, it is crucial for readers to have a comprehensive, well-connected understanding of the basics of networks and network effects in order to grasp all the implications of the analysis. In fact, before coming to this project, I myself thought I had a good understanding of networks and network effects. However, throughout my modeling and analysis of the Hardware–Software Game, I found myself continually researching the basics so as to understand what my analyses were telling me. What's more, much to my chagrin, I discovered that the resources available seemed to discuss the concepts surrounding networks and network effects in only a disconnected and disjointed manner. For these reasons, I provide a discussion in Chapter 1 of the features of networks, network effects, switching costs, and technology compatibility as they relate to the Hardware–Software Game. Readers who already have a good understanding of these concepts might want to skip to the last paragraphs of sections 1.1, 1.2, and 1.3 for a summary of the information discussed within each of these sections, or just read the summary at the beginning of section 1.4, which discusses the general role of network effects in the Hardware–Software Game.

As with networks and network effects, lifecycles play a crucial role in the Hardware–Software Game. Technology adoption lifecycles and lifecycles of network effects are both cause and effect, in that they both determine and are determined by the timing of new technology introduction by hardware providers and new technology adoption by software providers and users. As with the discussion of networks and network effects, readers who are familiar with the topics might want to skip to the summary paragraphs at the end of sections 2.1, 2.2, and 2.3 or just read the summary at the beginning of section 2.5, which discusses the general role of technology adoption lifecycles in the Hardware–Software Game.

Technology adoption lifecycles include the point of critical mass, known less formally as the *tipping point*. Readers familiar with technology adoption lifecycles will know that having new technologies achieve critical mass, in and of themselves, can be considered the definition of successful adoption. Given that the point of critical mass plays such an important role in technology adoption lifecycles, one would think that there would be much information in the literature about achieving critical mass. What I found during my research turned into the prototypical joke often told about economists assuming away any problems; that is, when discussing equilibrium points in the technology adoption process, economists note that there are two stable equilibria, the first corresponding to no adoption and the second to successful adoption, the latter of which entails achieving critical mass. Frustratingly enough, the economists then go on to say, "Assume the technology has reached critical mass . . ." and continue with the discussion. This handling of the adoption issue very conveniently skips right over one of the most essential aspects of the analysis, namely, how do you get your new technology to reach critical mass? Section 2.4 provides a discussion of critical mass dynamics, so readers unfamiliar with the subject can better understand its role in the analyses presented later in the book. Again, readers familiar with the subject might want to skip to the summary at the end of section 2.4 or at the beginning of section 2.5.

Chapter 2 ends with section 2.5, which discusses the general role of technology adoption lifecycles in the Hardware–Software Game.

The next two chapters, Chapter 3 and Chapter 4, discuss each of the players in the Hardware–Software Game, technology system users in Chapter 3 and hardware and software providers in Chapter 4. These chapters provide a discussion of what factors affect players' decisions to provide new technologies (in the case of hardware providers) or to adopt new technologies (in the cases of users and software providers) and how these factors change over time and in response to actions taken by the other players. I discuss the factors affecting users' decisions relating to new technology adoption in rather meticulous detail in Chapter 3, because it is the users who are the primary determinants of if and when hardware and software providers will end up being successful in the marketplace. Readers who are more familiar with user demand functions and how the size of the installed base, the stock of content, and hardware/software prices affect user demand for new system hardware might want to skim through sections 3.1.1, 3.1.2, and 3.1.3 quickly and/or skip to the summary paragraphs at the end of each of these sections. Likewise with user demand for new system content, users may want to skim through sections 3.2.1 and 3.2.2 quickly and/or skip to read the summary paragraphs at the end of each of these sections. However, even readers familiar with the topics should find the numerical examples in sections 3.1.4 and 3.2.3 useful for understanding how I apply my simulation model to specific scenarios.

Once the users' motivations are clearly understood, the hardware and software providers' motivations follow relatively easily and quickly in Chapter 4.

Chapter 5 gets into the meat of the book. It defines the structure of the Hardware–Software Game and analyzes its dynamics and outcomes. The analysis shows how the structure of the incentives faced by users and software providers (the chicken-and-egg problem) often leads to delays in adoption of new technology systems introduced by hardware innovators.

Chapter 6 provides readers with methods for addressing the delays in new technology adoption that result in many of the games. The chapter discusses general responses by hardware providers, together with scenario-specific means of speeding up the pace of adoption of their new systems.

Finally, Chapter 7 provides readers with a summary of key points from the analysis, tools to help readers apply the model to their own technology systems, and further areas of research that stem from the analysis.

For those who are interested, the underlying mathematical model I developed and used in the analysis is presented in Appendix A, and there is a list of further readings organized by topic in Appendix B.

CHAPTER ONE

Network Effects

When I was in graduate school, I used my Apple Macintosh computer to write papers and analyze data. However, after I finished school and got a job working at a consulting firm, I switched to a Dell PC. I would have preferred to stick with the Mac, because I knew all the ins and outs of working on the Mac and using Word and Excel. However, the business world, of which I was now a full-fledged member, used WordPerfect and Lotus. So, to be sure I could exchange information and documents with my colleagues and clients as easily as possible, it just made more sense for me to switch systems and use what everyone else in my new world was using.

This example illustrates how network effects can have a large impact on the incentives and actions of players in the hardware–software technology game. In particular, as a network becomes larger and more established, it becomes much more difficult for providers of alternative systems (Mac) to create enough value that users will switch from the established system (PC) to another one (Mac). Users will be more reluctant to switch from the established system when either of two conditions holds: (1) Users get value from being able to interact or share resources directly with other users of the same system (*direct network effects*), or (2) users get value from being connected to large systems because they get value indirectly from the large amount of associated resources, or the *ecology*, surrounding such systems (*indirect network effects*). In my case of being in the business world, both conditions held. I could exchange documents more seamlessly with other PC users, and if I needed a special program or gadget, my company's IT help person was much more likely to be able to get a PC version than a Mac version. At the same time, direct and indirect network effects also increase the costs of switching from one system to a different one. So, by increasing both the value of the technology system itself, as well as the costs of switching to a different system, network effects can effectively lock consumers into one system, to the exclusion of others. The easiest way for providers of new systems to get people to use their network (though not necessarily the way with the greatest expected profits) is to make their networks compatible with the established system. Of course, this also makes the current system even more entrenched.

This chapter provides a brief description of the aspects of network effects, switching costs, and compatibility that are important for understanding the Hardware Software Game. The first part of this chapter defines what network effects are, where they come from, and how they increase the value to users of being connected to a particular technology system. The chapter next discusses how network effects exacerbate the costs of changing networks, that is, how network effects can create lock-in for members of the system. Third, it considers how compatibility of concurrent or successive networks can be used to decrease the impact of system lock-in. The chapter ends with a discussion of how network effects influence incentives and actions of players in the Hardware–Software Game.

1.1 Definition and Sources of Network Effects

A *network* may be defined as a collection of objects, often referred to as *nodes*, or *members*, that are connected by *links* for purposes of exchanging information or sharing resources. Networks are especially common in the transportation and communications industries. For example, train stations (nodes) interconnected by train tracks (links) form a network, as do airports interconnected by airline routes, owners of telephones and fax machines interconnected by local and long-distance services, and owners of computers interconnected by both software systems and the Internet. In fact, most technologies that include both hardware and software or content components form a network.

If I showed you a network with four nodes and asked you how large the network was, you would probably be inclined to determine its size by the number of nodes and say that the network had a size of four. On the other hand, you might also determine the size of a network by counting the links between the nodes. For example, you could determine the size of the U.S. airline network either by counting the number of airports in the United States or by counting the number of direct routes between airports in the United States. In the latter case, the size of the network would grow as more direct routes are added, even if no new airports are built. And since people could get from place to place more quickly, I could say that the more interconnected network of airports would be more valuable than the less densely connected network.

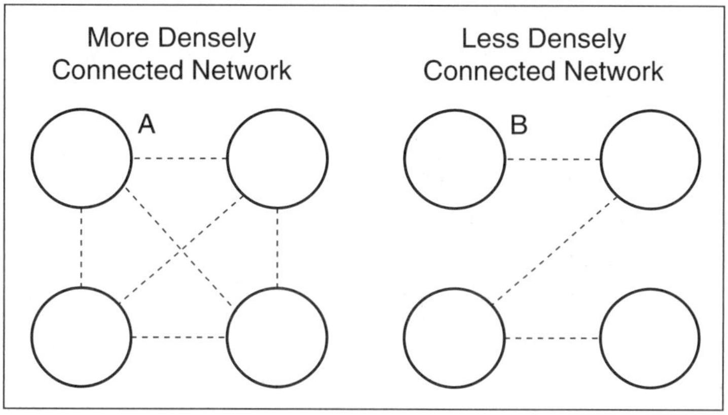

Figure 1-1 Networks

More generally, networks in which members are more densely connected, as in Network A in Figure 1-1, enable greater exchanges or flows of resources between members of the network than less densely connected networks, such as Network B. So, members of a network generally get more value from being members of more densely connected networks, even if the number of nodes is the same as in other, less densely connected networks.

At the same time, however, networks with more nodes enable more *potential* exchanges between members, even if all the members are not currently connected. In this case, networks with more nodes may be considered more valuable than networks with fewer nodes, even if the network with fewer nodes is more densely connected. This follows from the fact that it is usually relatively costly (in terms of money or resource requirements) to add a new node to a network, while it is generally less costly to add new links between existing nodes. In the end, then, it is generally the number of nodes, rather than links, in the network that determines its size, since it would be relatively easy to interconnect all the nodes. A network is said to exhibit *network effects* when the value of being a member of the network increases with the size of the network.

A network is said to exhibit *network effects* when the value of being a member of the network increases with the size of the network.

When networks exhibit network effects, the members should be willing to pay more to join larger (or denser) networks. However, this greater value associated with larger networks can cause dissension when the networks are created by technology systems that require the use of both hardware and software components from different suppliers. For example, when a network grows, say, from 100 members to 1,000 members, who gets to keep the extra value to the original 100 members created by the addition of the 900 new members? Do the hardware manufacturers get to keep this value? The software suppliers? Or do the members themselves get to keep this value? To some extent, whichever party is able to extract this extra value depends on whether the value is due to direct network effects or indirect network effects, which concepts are discussed next. More generally, though, this issue adds complexity to the Hardware–Software Game and will be discussed in more detail later on.

Believe it or not, I am generally slow to adopt new technologies. Even though I am fascinated by the evolution of technology and the phenomenon of adoption, I am generally reluctant to upgrade to the latest generation because of the time, energy, and money required to upgrade to the new version. And my sister is even slower to adopt than I am. I was relieved when she finally bought a cell phone, because she is not at home much, and I always had trouble getting in touch with her. So when she did buy a cell phone, not only did she enjoy the associated value by being able to call *me*, but she also created value for me because now I could call *her*. The new value she created for me is called a *network externality* or a *network effect*. An *externality* is the effect or impact that one person imposes on another when the first person does something without taking the other person into account. Network externalities arise because when people decide to become members of a network, they typically consider only the value to themselves of joining, but they do not consider the costs or benefits that their membership creates for existing members.

An *externality* is the effect or impact that one person imposes on another when the first person does something without taking the other person into account. Network externalities arise because when people decide to become members of a network, they typically consider only the value to themselves of joining, but they do not consider the costs or benefits that their membership creates for existing members.

Network externalities do not have to be positive. Negative externalities also exist, for example, when one more car joins a congested road during rush hour and causes all traffic to slow down. Another example of a negative externality is when your neighbors[1] decide to have a big, loud party on the night you are finally able to get to bed early.

The value created by a new member of a network can take one of two forms, either direct or indirect. New members of a network create value directly for existing members by enabling the existing members to link to the new members. *Direct network effects* are the added value created by the greater numbers of network members with whom existing members of the network can interact when new members join the network. So, for example, when my sister bought her cell phone, she created value for me by enabling me to call or text her wherever she is.

New members of an existing network can also create value for existing members of the network indirectly. As new members join the network, the size of the network grows and creates more profit opportunities for others who may want to provide accessory products or services to members of the network. *Indirect network effects* are the added value created by the increase in the availability of products and services associated with the increase in the size of the network. When my sister bought her cell phone, she made the installed base of users of our cell phone network larger and, therefore, more attractive to suppliers of cell phone services and accessories. This greater availability of accessories or services makes my cell phone more valuable to me.

In today's markets there are whole communities, or ecologies, of goods and services springing up around some core products. The growth of ecologies validates the core products and serves to make them more successful and desirable. An additional effect

> *Direct network effects* are the added value created by the greater numbers of network members with whom existing members of the network can interact when new members join the network.
>
> *Indirect network effects* are the added value created by the increase in the availability of products and services associated with the increase in the size of the network.

1. Neighborhoods and housing complexes form networks when they share the same infrastructure, such as roads, water systems, cable systems, and/or telephone systems.

of larger systems is that all parties become more and more interconnected to one another. In other words, once the size of the network reaches some threshold point, a positive feedback loop kicks in: The opportunities offered by members of the network attract providers of third-party goods and services, which makes the initial product more valuable to users, which, in turn, attracts more people to the network, which attracts more goods and service providers, and so on.

eBay provides a perfect example of the dynamics of indirect externalities. eBay recognized that many third parties, such as PayPal (before eBay bought it), were offering services to help eBay customers buy and sell products faster, more easily, and/or at better prices. The products and services offered by these third parties served to enhance the eBay experience and help eBay to become more successful. As a result, eBay eventually stepped in and began a company-sponsored initiative, the eBay Developers Program, to help these developers in their efforts to provide third-party products and services.

> *The eBay Developers Program is a program dedicated to helping 3rd parties integrate their businesses with the eBay Platform. As a member of the Developers Program, you can create custom applications that effectively and efficiently handle almost all types of business transactions on eBay.*[2]

According to a news release from November 2005, "The eBay Developers Program has more than doubled over the past year, to 21,000 members and more than 1,600 member-developed applications,"[3] which help to support an eBay community of buyers and sellers totaling more than 100 million people around the world.[4]

To summarize, a central feature of networks is that as the size of the network expands, the value to existing members of being a member of the network increases, or in other words, networks exhibit *network externalities* or *network effects*. The value added to existing users when new users join a network may take the form of either direct or indirect network effects. *Direct network effects* occur when larger networks create more value than smaller networks by enabling members of the larger networks to interact with more members of that network. *Indirect network*

2. http://developer.ebay.com/faq/whatisdevprog.
3. http://developer.ebay.com/programs/marketing/freeaccesspr/.
4. http://pages.ebay.com/aboutebay/community.html.

effects occur when the consumption experience associated with larger networks is enhanced by the greater availability of complementary products and services associated with the larger networks. Readers interested in further information on networks and network effects can find a list of resources in Appendix B.

1.2 Switching Costs

My dad was an eye doctor, specializing in the care of children. He performed surgery on some of his patients, straightening out their crossed eyes. His patients were his family, and his patient charts were like scrapbooks of their lives. The files contained notes in the margins with personal information about his patients—where they went to school, what subjects they liked best, whom they married, whether they were having any problems with family or work—so he could treat them as *people* rather than patients. Also, whenever he performed surgery, he would take before and after Polaroid pictures, which he taped to the inside of the patients' folders. Furthermore, as part of the medical information he recorded, he would draw pictures of the patients' eyes, so he would have a more accurate spatial record of where the problems were. The files additionally included hospital and surgery records, correspondence with other doctors, insurance and payment information, and, for a surprisingly large number of patients, thank-you cards and drawings that the patients had sent my dad.

When computers had finally taken hold in the late 1980s, and especially during the early 1990s, my dad grappled with whether or not to switch his medical records from paper to electronic files. The changeover would have taken at least several weeks and would have entailed many steps: purchasing the appropriate software, along with additional computers; hiring people to transcribe the information in the medical files into electronic data files; and purchasing a digital camera to replace the Polaroid pictures with digital photos. At the time, computer memory was expensive, and computers were relatively slow, so storing the files and pictures electronically would have been costly, and uploading and downloading files time-consuming. He would have needed to buy a scanner to create digital files from the paper insurance information, hospital records, and correspondence, since at that time very few, if any, of the parties in the medical industry exchanged information electronically. He would have also had to dispose of or continue to store the paper records. Finally, he and his staff would have had to spend many hours learning to use the new system, and during the transition it would have taken him much longer to enter the information into computers that he used to scribble (in very bad doctor's handwriting) quickly onto the charts. And what if, after the new system had been installed, having everything done digitally turned out to be no cheaper or easier than simply using paper, especially since relatively few other doctors in his network were switching to digital? In the end, my dad decided that the time and money costs of switching from a paper office to a digital office at that time were just not worth it.

Switching costs associated with moving from old technology systems to new ones are all the costs consumers must bear when they leave their old systems and suppliers and establish themselves with new ones. As in the case of my father's medical records, switching costs may be plentiful. They can include, for example, any penalties or fees users must pay to break contracts (such as leases) or end agreements with current suppliers to free themselves up to move on to new ones; uninstallation and disposal costs required to remove old system components; equipment and installation costs required to install new systems; time and money costs required to train users to operate new systems; costs associated with downtime while transitioning from old systems and ramping up to obtaining the full benefits of new systems; any costs associated with losing connectivity to users or content that remains on old networks; any risks associated with new systems not living up to expectations, not providing all the expected benefits, ending up requiring more than the expected time or money costs to use, and so on.

> *Switching costs* associated with switching from old technology systems to new ones are all the costs consumers must bear when they leave their old systems and suppliers and establish themselves with new ones.

For it to be worthwhile for users to switch to a new system, the benefits associated with the new system must outweigh not only the costs of purchasing the hardware and software for the new system, but also the costs associated with transitioning from the old system to the new one, as seen in Figure 1-2.

In other words, when switching costs are high, the benefits of the new system must be much greater than the benefits of the current system to make switching an attractive proposition. In particular, for markets in which network effects are strong, switching

Potential User Chooses to Adopt New Technology System If
Benefits of New Technology System >
[Benefits of Old Technology System
+ Costs of New Technology System
+ Switching Costs]

Figure 1-2

costs associated with downtime during the transition from one network to another, and the costs associated with losing connectivity to users or content that remains on the old network, can be particularly large. In these cases, the benefits of a new technology system must be a whole lot better than those offered by the current system to get users to move out of their current networks and into new ones.

A good example of switching costs involves telephones. Historically, people were unable to keep their same telephone number when they switched from one service provider to another. If they decided to switch telephone service providers, telephone users had to give up their current telephone numbers and get new ones when they established service with new providers. This made it especially costly for people who relied heavily on being contacted by their many friends, relatives, colleagues, and/or business associates to switch telephone service providers. Recognizing that this greatly impeded competition in the telephone service markets, Congress changed the Telecommunications Act in 1996 to allow for *telephone number portability* when customers switched providers. In other words, instead of having to give up their telephone numbers when they left old service providers, telephone users could take their telephone numbers with them and use the same numbers when they signed up with new service providers. After much resistance by the cellular service providers, telephone number portability was extended to the cellular telephone market and became fully implemented in May 2004, so as to foster competition among cellular service providers.

Since the presence of switching costs makes it more difficult for new technology systems to draw customers away from current technology systems, it is not surprising that many existing suppliers try to impose large cancellation fees and do whatever else they can to prevent users from switching from their system to a different one. Such *lock-in* tactics also enable suppliers to charge users higher prices than they might otherwise, because despite the higher prices, the switching costs still make it too expensive for many users to switch.

MCI's "Friends and Family" program provides a novel example of how a provider created lock-in for its customers. The way the promotion worked was that customers of MCI's long-distance telephone service would designate lists of people whom they wanted in their own friends and family networks. MCI would then call upon the designees to switch to MCI's service, in which case both users would receive a discount for all phone calls they made to each other. This program had two interesting effects. First, users found it in their interest to recruit their friends and family to switch to MCI's service so they could receive discounts on their phone calls. This enabled MCI to transfer some of the burden of trying to get people to switch to their network onto current users. Second, MCI's "Friends and Family" program helped entrench its users into the network; that is, once a network of friends and family was established, no one user had an incentive to unilaterally switch to another service provider, since he or she would then lose the discounts on phone service to the self-proclaimed network of important contacts.

To summarize, *switching costs* encompass any costs or investments consumers must bear when they leave their old products and suppliers and establish themselves with new ones. When switching costs are high, switching to new systems will be worthwhile only when the benefits offered by new systems are significantly greater than the benefits provided by current systems. In particular, over time, as people have become more connected to each other through networks, users' switching costs of moving from one network to another have generally increased. This increasing interconnectedness—that is, greater direct and/or indirect benefits associated with being members of particular networks—has generally made it more difficult to get new technology systems adopted, and even more so when they are not compatible with current networks. In the Hardware–Software Game, switching costs make it more difficult for hardware manufacturers to get users and content providers to switch from currently established systems to new technology systems. Readers interested in further information on lock-in and path dependence can find a list of resources in Appendix B.

1.3 Compatibility

Sony and Philips Consumer Electronics jointly created the compact disc (CD). The CD was originally developed for storing digital audio. When CDs came onto the market in 1982, their higher-quality audio and greater durability and portability were meant to lead audio listeners to replace vinyl albums with CDs. CDs represented a newer technology generation that was incompatible with the earlier-generation vinyl albums. In other words, a CD player cannot also play vinyl albums.

In 1992 Dolby Labs introduced Dolby Digital, an audio format that "was developed specifically for multichannel applications, including film sound and digital surround sound in the home."[5] In 1993, DTS, Inc., introduced an alternative audio format, DTS (Digital Theater System):

> *DTS created a format that makes audio tracks more dynamic, more realistic and more closely matching the original audio master than other digitally encoded soundtracks. Coupled with the multi-dimensional benefit of surround sound technology, the audio quality dramatically improves and enhances content.*[6]

Dolby Digital and DTS use the same-generation technology, and they are both used on CDs and DVDs. However, the two formats are incompatible with each other: Audio systems must specifically be Dolby Digital–compatible or DTS-compatible before they are able to play the designated format. Dolby Digital is standard on all DVD-video movies for home theaters, whereas DTS is optional.[7]

More generally, when a new innovation is brought to market, it is usually meant either (1) to offer a more evolved (next-generation) alternative to what is currently available, such as updating vinyl albums with CDs, or (2) to offer a similar (same-generation) alternative to what is currently available, such as Dolby Digital versus DTS. In this sense, then, similar technologies may be *compatible* or *incompatible* in either of these two respects.

In the first case, products may be compatible or incompatible with the products they are meant to replace. If they are compatible with technology from previous generations, then the new technology system will work with the hardware and/or software from the older technology system, which generally means that the new hardware will

5. www.dolby.com/about/who_we_are/history_5.html.

6. www.us.design-reuse.com/exit/?url=http://www.dts.com.

7. www.timefordvd.com/tutorial/SurroundSound.shtml.

be able to access the older content/software. Conversely, if they are incompatible, the new stuff will not work with the old stuff.

In the second case, similar technologies are compatible with other technology systems from the same generation if they can be used interchangeably in contributing to the user's consumption experience. For example, Sony and Panasonic CD players are compatible in the sense that they can be used interchangeably in an entertainment system to play audio CDs. A more interesting example is the case of incompatible cellular telephones that can access only one broadcast standard, either CDMA in the United States or GSM in Europe. Compare this with compatible, dual-mode phones that can be switched to access either CDMA signals for use in the United States or GSM signals when the user is in Europe.[8]

A tremendous amount has already been written about the ins and outs of technology compatibility. I do not plan to rehash what has already been done. Rather, I

> Technology systems are *compatible with technology from previous generations* when the new technology system will work with the hardware and/or software from the older technology system.
>
> Technology systems are *compatible with other technology systems from the same generation* if they can be used interchangeably in contributing to the user's consumption experience.

8. The evolution of cellular standards in the United States and Europe simultaneously provides examples of all four possibilities of compatible versus incompatible, next-generation versus same-generation technologies. In the United States, the first-generation (1G) cellular communication system, analog, was generated without competition by AT&T. Europe used a similar 1G system.

Early on, the European Telecommunications Standards Institute (ETSI) standardized second-generation (2G) cellular service in Europe on the GSM system, which had been created by a group of European handset, chip, and infrastructure firms, led by Ericsson, to enable better integration of the mobile infrastructure and handset system throughout Europe. In the Unites States, development of the 2G system was left open to competition. Initially, TDMA, established by Ericsson, became the de facto standard, though GSM service was also used in the United States. Soon afterward, however, Qualcomm entered the market with its CDMA technology, and a ferocious battle ensued. For a while, the three standards, TDMA, GSM, and CDMA, simultaneously coexisted, but Qualcomm's CDMA eventually triumphed over TDMA, leaving only CDMA (used by Verizon and Sprint) and GSM (used by Cingular).

For third-generation (3G) systems, the United States has an extension of CDMA, CDMA2000, and Europe, recognizing the superiority of Qualcomm's technology, designed a version called WCDMA.

briefly discuss the basics of technology compatibility so the reader has some understanding of the role it plays in technology adoption and how it affects the Hardware–Software Game.

By making concurrent or successive technology systems compatible with one another, new system innovators may be able to moderate some of the lock-in problems that might otherwise deter users from leaving their old systems to adopt new ones; that is, system compatibility can help to ease the transition from one technology system to another. In addition to this benefit, however, each aspect of technology compatibility has other upsides and downsides.

The downside of making next-generation technology systems backward-compatible is that the dual functionality means that the new system will generally be more complex, and perhaps also bulkier and more expensive, than the associated stand-alone system. An additional disadvantage for manufacturers of having backward-compatible systems is that it tends to slow down the rate of adoption of the newer-generation software or services. When users can continue to use their old content, there is less urgency to upgrade to the new content.

As for compatibility of simultaneous generations, by making one's technology compatible with that of other currently available systems, an innovator essentially trades a large slice of a small pie for a small slice of a larger pie. Think Apple versus Microsoft.

As it applies to the analysis at hand, one clear benefit of compatible same-generation systems is that when there are network effects, having one large, compatible market will generally create the potential for more total value than having two smaller, incompatible markets, as illustrated in Figure 1-3. This reflects the concept of network size, discussed in section 1.1, as the total potential number of links between members of a network. As seen in Figure 1-3, with one large network (of six nodes in the figure), there are more potential links between members (15 total links) than there are with two smaller networks (with two small networks of three nodes each, there is a total of six potential links). In this sense, then, there is a *multiplier effect* associated with networks, where the total potential value associated with a collection of members is greater than the sum of the values associated with each individual member.

Another benefit of standardization is that achieving critical mass (there is a detailed discussion of critical mass in the next chapter) is usually easier, since large-scale adoption is much more likely once the technology becomes functional. This is especially important for new innovators that do not have an established reputation in the marketplace to help speed the pace of adoption of their new technology systems.

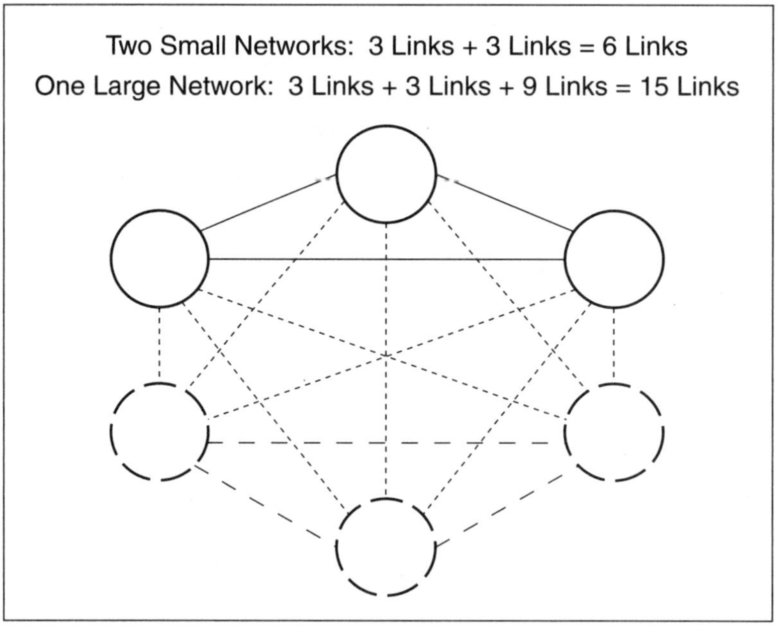

Figure 1-3 Total links: large versus small networks

To summarize, similar technologies may be *compatible* or *incompatible* in either of two respects: (1) with successive generations of products, or (2) with the same generation of products. In the first case, compatibility of technology systems means the new hardware can access the old software or other content. In the second case, similar technologies are compatible if they can be used interchangeably in contributing to the user's consumption experience.

The main advantage of compatibility of successive technology generations is that it speeds up the adoption of next-generation hardware by reducing the risks and costs to users of upgrading. The big disadvantage is that compatible technologies tend to be more complex, bulkier, and/or more expensive than incompatible technologies.

Product compatibility of same-generation products generally entails sacrificing a large slice of a smaller pie in exchange for a small slice of a larger pie. Suppliers tend to prefer product compatibility—that is, a smaller slice of a larger pie—(1) when network effects are strong, in which case the total size of the pie will be much larger with compatibility; in this case, though, users face the potential problem of becoming more entrenched in the larger network; and (2) when achieving critical mass is a major concern, such as when suppliers are small or have not yet generated a reputation in the marketplace; in this case product compatibility mitigates, if not totally eliminates, this problem. Readers interested in further information on standardization and compatibility can find a list of resources in Appendix B.

1.4 Network Effects and the Hardware–Software Game

In this chapter I have discussed how the value of networks increases with size. These *network effects* can come in either of two forms, direct or indirect. *Direct network effects* occur when larger networks provide members with more value by enabling them to communicate or exchange resources with more members of the network. *Indirect network effects* occur when larger networks lead to greater availability of products and services provided by third-party suppliers, thereby increasing the value of being a member of the network. Next, I discussed how the value associated with direct and indirect network effects can end up locking members into a technology system, either by increasing the value to members of staying with that system, or by increasing the costs to members of switching to another technology system. Finally, I discussed how compatibility can be used by system innovators to mitigate the effects of lock-in.

Let's now return to the Hardware–Software Game and consider how network effects shape the incentives of the game's players and the eventual outcome of the game.

When technology systems exhibit network effects, part of the value of these systems to users comes from (1) being connected to other members of the system, in the case of direct effects, and/or (2) having a lot of content available, in the case of indirect effects. If the value to users of being connected to that network constitutes a large part of the value of the technology system, then users will be reluctant to change systems if it means they lose that connectivity. In this case, new system providers force users to choose between the prospect of adopting new technology systems with few or no established networks, or staying with old systems with large, established networks. In these cases, users will tend to choose to stay with old systems, unless the added value promised by the new system is substantial. So innovators of new systems face the challenge of creating the large value necessary to pry users out of existing systems and convince them to switch to new ones.

Consider now the issue facing software providers. They are currently supplying content to large, established technology systems. The substantial sizes of the systems and their associated network effects make it profitable for software providers to supply these markets. Innovators come along and provide frameworks for new technology systems. Software providers then face the prospect of continuing to dedicate their resources to the provision of content for currently established, proven-to-be profitable networks, versus switching their resources to the provision of content for new technology systems whose networks may or may not eventually become established. Clearly, innovators of new systems will have a difficult time convincing software providers to leave their secure, profitable environments for new, risky ones.

It follows that the greater the extent to which the value of currently established systems is heavily tied to the networks associated with those systems, the greater the difficulty new system innovators will have in convincing users and software providers to disconnect from old systems and establish themselves with new ones. In these cases, overcoming this reluctance to leave old systems for new ones, thereby winning the Hardware–Software Game, will require system innovators to minimize the costs and/or to maximize the benefits to both users and software providers of switching from established systems to new ones. Solutions to these problems are discussed in more detail in Chapter 6.

Technology Adoption Lifecycles

> *For the first three millennia or so since the invention of paper, the writing instrument of choice in Western . . . culture was the quill. One simply found a goose who could be persuaded to donate a tail feather, harvested the feather, allowed it to dry . . . "gutted" it and trimmed off the excess fuzz. Then, after softening up the quill under moist heat, one used a pen-knife . . . to shape the tough, horny shaft into a good writing point and split it to hold a small amount of ink. Then, one dipped the quill in ink and wrote a line, dipped, wrote a line, dipped . . . The first step toward turning the pen from a handmade tool into a manufactured commodity was taken sometime in the early 19th century, when mass-produced steel pen points began to appear.*[1]

Nineteenth- and twentieth-century innovators in the designing of writing instruments grappled with the problems of creating pens that provided smooth flows of ink, without skipping or blotting, and that were easily and cleanly filled with ink.

> *The period from 1925 to 1940 is regarded as the zenith of the fountain pen era, with many beautiful and highly functional writing instruments appearing all over the world . . . With the development of the metal point, the feed, and the various self-filling mechanisms, [during this period] we at last had the basis of the modern, portable, self-contained fountain pen.*[2]

After decades in the making, the ballpoint pen was introduced in 1945, but it was not until 1952 that Marcel Bich "was ready to introduce his new wonder: a clear-barreled, smooth-writing, non-leaky, inexpensive ballpoint pen he called the 'Ballpoint Bic.' The ballpoint pen had finally become a practical writing instrument."[3]

1. www.rickconner.net/penspotters/history.html.
2. Ibid.
3. www.ideafinder.com/history/inventions/ballpen.htm.

Technology adoption lifecycle refers to the changing nature of producer and consumer attitudes and actions, from the early stages of a technology system's introduction into the marketplace, through its early adoption, its establishment as a market presence, and finally its waning stages before it becomes fully superseded by an alternative system.

The Hardware–Software Game generally takes place as producers and consumers move from one system that is in the later stages of its lifecycle to an alternative system that is in the earlier stages of its lifecycle. Also, as will be seen in this chapter, the incentives and actions of producers and consumers differ depending on which lifecycle stage a technology is in. It therefore follows that to understand how the Hardware–Software Game for a new proposed technology system will play out, the nature of the lifecycle stages of both the old technology system and the new one that proposes to replace it must first be understood.

This chapter first introduces production and consumption lifecycles, a discussion that takes us through the progression of five stages of technology production and four adopter groups. Second, the chapter considers lifecycles of network effects, and the changing nature of the value to users provided by direct and indirect network effects, as the technology adoption lifecycle progresses. Third, the chapter emphasizes how the adoption of a new technology system depends on the incremental value it provides over existing systems. Fourth comes the point of critical mass and the nature of direct and indirect network effects during the key stages of a new system's adoption. Finally, the chapter ends with a discussion of the role of technology adoption lifecycles in the Hardware–Software Game.

2.1 Production and Consumption Lifecycles

I have a brother who is a real technology maven. When we were little, he used to have disemboweled clocks and radios and other such gadgets lying around his room. I'm sure he must have spent hours and hours investigating and tinkering and learning, but I didn't see any of that. All I know is that he just seemed to know how everything worked. When desktop computers first came out, he latched on to them and has been seated in front of one ever since. He owns the latest generation of computers and cameras, he has more gadgets than Brookstone, and he always seems to know what's coming around the corner technology-wise before anyone else. He's the epitome of an innovator, a true lover of technology.

Everett Rogers was a pioneer scholar in the area of new technology adoption and diffusion. The following discussions of adopter groups are taken from his seminal work, *Diffusion of Innovations,* and illustrated in Figure 2-1.[4]

More generally, *innovators* are characterized by their "ability to understand and apply complex knowledge." They are "venturesome" individuals, and their interests and attitudes tend to set them a bit apart from the larger society. Yet, it is these users who are responsible for "launching new ideas into the system."

Early adopters comprise a more mainstream portion of society than innovators. Early adopters are attracted to new gadgets that offer never-before-seen features and capabilities, and they enjoy playing with them, seeing how they work, and imagining all their possible uses. They do not care that all the bugs have not yet been worked out, nor do they care that a whole lot of people do not yet have the device. In fact, part of the initial attraction is that early users are an exclusive bunch, the first to know about new advances and to access new innovations. Since early adopters have the "respect" of their peers, it is the early adopters who "help trigger the critical mass when they adopt an innovation."

For those technologies that survive the early adoption process, manufacturers have been able to work out the bugs, and through early adopters they have been able to substantiate the benefits of their new products. During this process, their technologies have been moving into the second phase of their lifecycles. Manufacturers are beginning to understand how to make the product better and faster, as they benefit from learning by doing and economies of scale and specialization in the production process. In other words, as they ramp up production, manufacturing costs have been dropping, so the product is becoming not only better, but also less expensive.

At this point, the rank of adopters has passed through the small group of early adopters and moved on to the *early majority*. These new users are more "deliberate" than early adopters in deciding whether or not to adopt a new product. They take their time investigating the product, probing the early adopters for tales of their experiences, testing the product for themselves, and making sure the bugs have been worked out and that they can get the product for a more reasonable price. This is the heart of the adoption process. Interpersonal communication networks between peers kick in, word spreads, validation is achieved, and adoption takes off.

4. Rogers 2003, 282–85.

As the early majority starts to adopt the new technology, other manufacturers see that there are profits to be made. They enter the market with products that both substitute for and complement the original technology product, which increases competition. The new rivals try to beat out each other's products and win market share by making the product ever better (through product innovations) and cheaper (through process innovations). By the time the bulk of the early majority has purchased the new innovation, the product starts to become commoditized; that is, rather than focusing on significant aspects of product improvement, manufacturers focus on making the product at lower costs, which leads to lower prices. By this point, the original product innovators have often refocused their efforts on finding new applications for the product or evolving the product into something more, rather than simply trying to sell their original product to latecomers.

After the early majority, the next group to adopt a new innovation is the *late majority*. These folks are a "skeptical" lot, and so most of the risks associated with product adoption, such as whether or not the product will work as advertised, must be resolved before the late majority will be convinced to step up. The last group to adopt a new innovation is the *laggards*. They are "traditionalists," are generally suspicious of change, and adopt new technologies only when absolutely necessary.

This basic description of the *production lifecycle* of a new innovation, as illustrated in Figure 2-1, thus encompasses an early stage of production, product takeoff, product competition, product commoditization, and the late stages of product manufacturing.

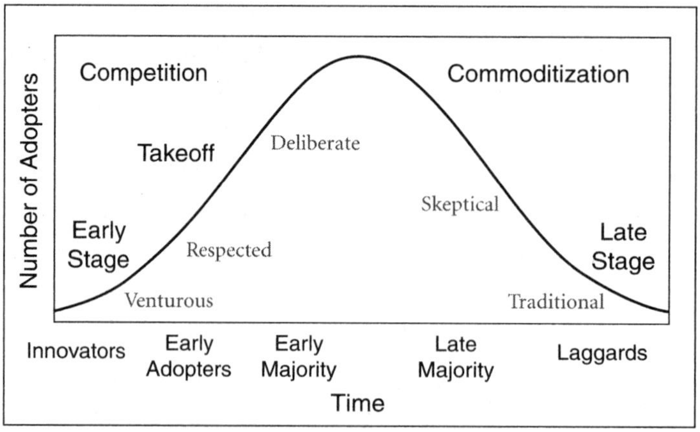

Figure 2-1 Technology adoption lifecycle

At the same time, the *consumption lifecycle* of a new innovation entails adoption first by innovators and early adopters, followed by the early, then late, majorities, and finally the laggards.

To summarize, the adoption process of new products by the marketplace has associated production and consumption lifecycles. The *production lifecycle* of a new innovation encompasses five stages: (1) an early stage of production, during which time the product may still contain bugs and is costly to produce; (2) product takeoff, during which time most of the bugs have been worked out and production costs decrease, due to economies of scale and specialization; (3) product competition, during which time new suppliers of the product enter the market and start to compete with one another for sales; (4) product commoditization, during which time the focus of competition in supply of the product transitions from product novelty to product price; and (5) the late stages of product manufacturing, during which time the product has become fully commoditized and new generations of the product are being introduced.

At the same time, the consumption lifecycle of a new innovation entails adoption by four general groups of users: (1) innovators and early adopters, risk takers, who are attracted to novel innovations that offer new and different features and capabilities; (2) the early majority, who are more deliberate in their purchasing decisions, requiring bug-free products whose value has been validated by early adopters; (3) the late majority, a skeptical lot, who demand low prices and large amounts of product support; and finally (4) laggards, the traditionalists, who adopt new innovations only when forced to do so. Readers interested in further information on new technology innovation and adoption can find a list of resources in Appendix B.

2.2 Lifecycles of Network Effects

Palm characterizes its basic handhelds as "your planner, journal, sticky notes, and calendar all in one little place."[5] In other words, if you buy one of their personal digital assistants (PDAs), you immediately have access to an electronic address book, calendar,

5. www.palm.com/us/products/handhelds/z22/index.html.

notepad, alarm clock, and so on. Additionally, Palm also offers supplementary software for its PDAs, including various games, dictionaries, spreadsheet programs, and database programs, among others, together with various accessories, such as cases, adapters, and headsets.[6]

The case of the Palm PDA illustrates how the total value of a technology system that exhibits network effects has two separate parts. The first part of the system's total value is what I refer to as *technology value.* This is the value users place on having a PDA that gives them access to the basic functionality—that is, an electronic address book, calendar, notepad, alarm clock, and so on—plus the value of having the option of being able to buy supplementary software and accessories, so as to enhance the basic offerings. More generally, technology value is the value users get from being able to achieve or unlock the consumption experience provided by the basic technology plus the value of being able to access software or other content that is currently available or that users expect to become available in the future for the new technology system. This latter component is an option value, if you will, on being able to access the content upon demand.

> *Technology value* is the value users get from being able to achieve or unlock the consumption experience provided by the basic technology plus the value of being able to access software or other content that is currently available or that users expect to become available in the future for the new technology system.

The second part of the system value of a PDA to users includes two sub-parts. The first sub-part is the value users get by being able to exchange information on the spot with other owners of PDAs. This value of being able to interact or otherwise share resources with other users is the (direct) value associated with the PDA's *direct network effects.* The second sub-part is the value users get by being able to choose from a wide variety of software and applications for their PDAs. These complementary components may be provided either by the original manufacturer or by some other third parties

6. http://us.store.gc.palm.com/.

that have been attracted by the large size of the network. The value generated by these complementary components is the value associated with the PDA's *indirect network effects*.

There is one final *indirect value* associated with having a large installed base of users, which is a measure of validation for the new technology system.[7] More specifically, when a new system is introduced into the marketplace, there is a risk to users of adopting the new system, because the technology may not be all that the manufacturer is claiming it to be. But as more and more users adopt the new system, these adopters become a signal to the rest of the community that the technology is in fact worthwhile. This signal serves to decrease the risk to new users that they will be left holding a dud if they adopt. As a case in point, Motorola's head of marketing for mobile devices had this to say about their cell phones: "The strongest marketing tool is the first 20,000 people who buy the device. If they like it, they will tell their friends."[8]

> The *direct value* associated with *direct network effects* is the value users get from being able to interact or share resources with other users.
>
> The *indirect value* associated with *direct network effects* is the validation value that early adopters create for subsequent adopters.
>
> The *value* associated with *indirect network effects* is the value users get from having access to a wide variety of content.

Finally, then, I have the total value of a technology system as being composed of four components, as illustrated in Figure 2-2: (1) the value of the technology itself; (2) the value of direct network effects associated with the installed base of new system users, with whom members of the network can interact; (3) the value of "indirect" direct network effects of the installed base, as a measure of validation of the technology system; and (4) the value of indirect network effects associated with the availability of software or content to enhance the consumption experience provided by the technology system.

---

7. See, for example, Bass 1969.

8. Holson 2008.

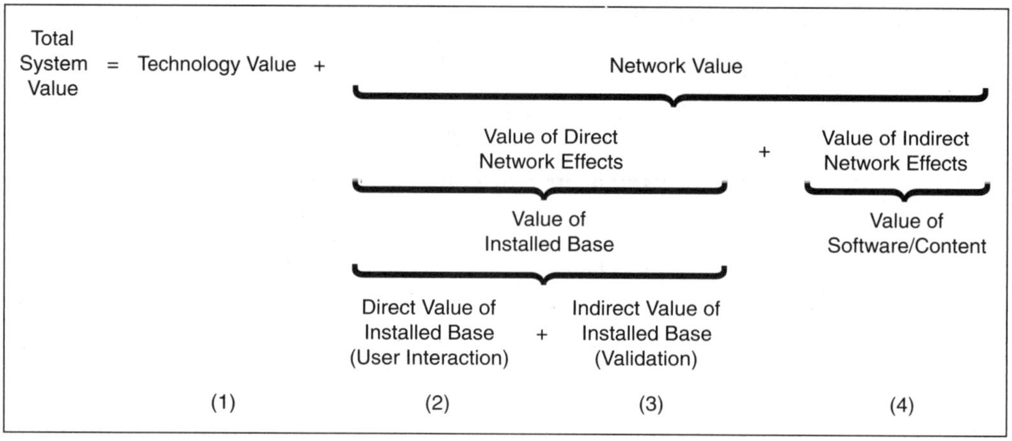

Figure 2-2

For potential users to decide to adopt the new system, they must believe that the total value of the system is worth the costs, where costs include both the price of the technology as well as any switching costs (described in section 1.2), as illustrated in Figure 2-3.

Now let's think about how these various components of value change over time. When new technologies first come onto the market, by definition, there are not a lot of people who have yet adopted them. So at the early stages of adoption, any direct and/or indirect network externalities or effects will be small.

At the same time, when technologies are new, the costs of making them are high, because manufacturers have not yet been able to benefit from learning by doing or economies of scale. Since the costs are high, the prices users will have to pay will generally be high. Furthermore, the total costs of technology adoption include switching costs. Switching costs for early adopters tend to be high, since they include the risks that as-of-yet-unvalidated technologies will not function as claimed.

Potential User Chooses to Adopt If

Technology Value + Network Value > Total Cost of Technology

Figure 2-3

For Early Adopters

Technology Value + Network Value > Total Cost of Technology
(Low) (High)

=> Technology Value Is High

Figure 2-4

So now we see that during the early stages of product introduction into the marketplace, the price is high and the network value is low. It follows that for the total value of the technology to be worth the price to early adopters, the value they place on the technology itself must be high, as illustrated in Figure 2-4.

At the other end of the adoption lifecycle, a relatively large number of people have bought the technology—those people with high and moderate technology values. The holdouts must therefore have very low technology values; otherwise, they would have adopted the new technology by now. At the same time, manufacturers have been able to move down their cost curves by taking advantage of learning by doing and economies of scale. The cost of making the technology will thus have dropped considerably from the time it was first introduced. It follows that the late majority must place a large value on the networks associated with new technology systems, as illustrated in Figure 2-5.

So now I have established that during the early stages of adoption, those who adopt new technologies will have high technology values and low network values, whereas those who adopt during the later stages of technology lifecycles will have low technology values and high network values. This relationship is illustrated in Figure 2-6.

This description should make sense when you think about the characteristics described previously of early versus late adopters of new technology. Early adopters like technology for technology's sake, and they are risk takers, so they can accept small

For Late Adopters

Technology Value + Network Value > Total Cost of Technology
(Low) (Moderate)

=> Network Value Is High

Figure 2-5

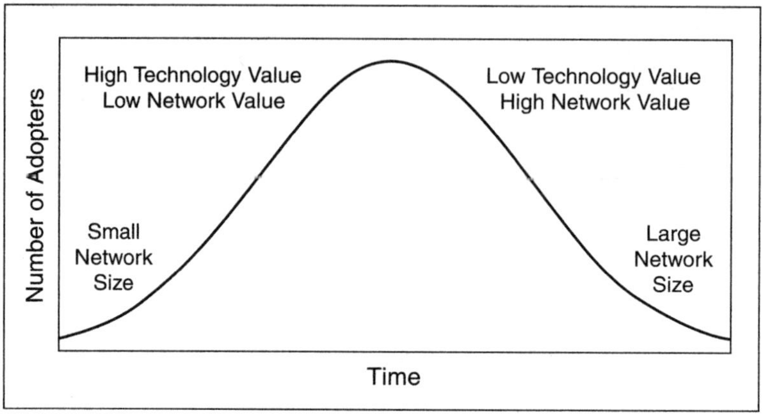

Figure 2-6 Lifecycle of technology and network values

imperfections in how the technology works. They do not need to have a large amount of content available to enjoy the benefits the new technology has to offer. It is thus perfectly consistent for them to have high technology values and low network values. Late adopters, on the other hand, do not like taking risks or dealing with imperfections. By the time they adopt, the risks of malfunction have all but disappeared, and there is plenty of added content, as well as accessories and services, to enhance the value of the technology for them. It is thus consistent for them to have low technology values and high network values.

> Earlier adopters have higher technology values and lower network values, and later adopters have lower technology values and higher network values.

I shall now take a slightly different perspective. I have been viewing the technology adoption lifecycle as a series of static stages: early stage, takeoff, competition, commoditization, and late stage. I will now, instead, think about it as a continuous dynamic process. The first users to adopt the new technology have the highest technology values of all potential users and the last to adopt have the lowest. So then throughout the lifecycle of the adoption process, the technology value of the new innovation will fall, as illustrated in Figure 2-7.

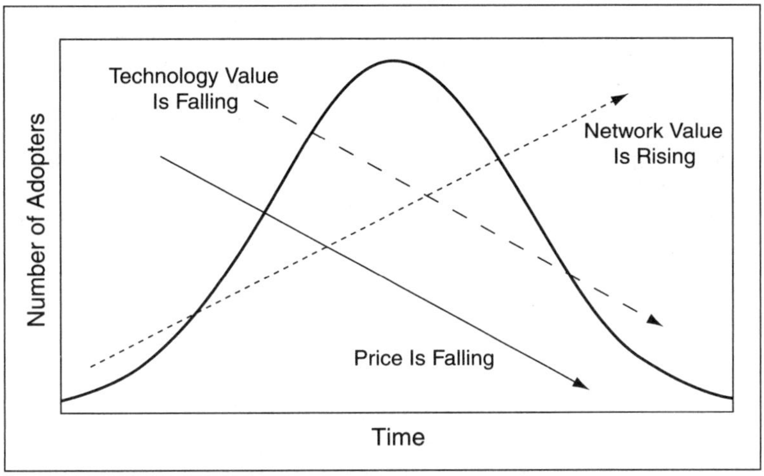

Figure 2-7 Technology adoption lifecycle dynamics

Similarly, the network value of a new technology starts out low during early stages of adoption and ends up high during late stages of adoption. Network value must therefore rise throughout the technology adoption lifecycle. Finally, the price of a new technology starts out high and falls throughout the adoption lifecycle.

> Throughout the lifecycle of the adoption process of a new system,
>
> - Technology value decreases
> - Network value increases
> - Price decreases

To summarize, I established that the total value of a technology is equal to the sum of the value of the technology itself, plus the value of the direct and indirect network effects associated with the new technology. Users will choose to adopt the new technology when the total technology plus network values of the technology exceed its total cost to users. I also established that at the early stages of a network product's adoption lifecycle, the technology value to users is high, and it decreases throughout the adoption lifecycle as the nature of adopters changes; the network value to users starts out low, and it increases throughout the adoption lifecycle as complementary products and services become increasingly available; and the price starts out high and decreases throughout the adoption lifecycle with the realization of economies of scale, specialization, and other forces of competition. Readers interested in further information on networks and network effects can find a list of resources in Appendix B.

2.3 Technology Replacement Lifecycles

"On July 26, 1775, members of the Second Continental Congress, meeting in Philadelphia, agreed

> *That a postmaster General be appointed for the United Colonies, who shall hold his office at Philada, and shall be allowed a salary of 1000 dollars per an: for himself, and 340 dollars per an: for a secretary and Comptroller, with power to appoint such, and so many deputies as to him may seem proper and necessary.*
>
> *That a line of posts be appointed under the direction of the postmaster general, from Falmouth in New England to Savannah in Georgia, with as many cross posts as he shall think fit.*

"This simple statement signaled the birth of the Post Office Department, the predecessor of the United States Postal Service and the second-oldest federal department or agency of the United States of America."[9]

"What hath God wrought" was the message that introduced the use of the electrical telegraph in 1844, which had been developed and patented by Samuel Morse.[10] The commercialization of the telegraph, in large part by Western Union, replaced much of the communications previously handled by the post office.

Alexander Graham Bell received a patent for the telephone in 1876; in 1877 the Bell Telephone Company was formed, and in 1878 the first commercial switchboard

9. www.usps.com/cpim/ftp/pubs/pub100/pub100_001.htm. The Bureau of Indian Affairs is the oldest federal department in the United States.

10. http://inventors.about.com/library/inventors/bltelegraph.htm.

began operation.[11] The telephone soon replaced much of the communications previously handled by the telegraph. Cordless telephones replaced corded telephones during the mid-1980s; and wireless telephones had replaced most cordless phones by the late 1990s.

This sequence illustrates how evolutionary and revolutionary new technologies replace older technologies that had previously been used to achieve the same or similar goals, albeit perhaps less efficiently or effectively. The point is that when a new technology system is introduced into the marketplace, there is almost always an existing alternative for users to accomplish the same thing that the new technology system has to offer. Said differently, when users adopt a new technology system, they generally use the new technology system to replace an existing technology system.

> When users adopt a new technology system, they generally use the new technology system to replace an existing technology system.

My television set is a 32-inch ProScan (RCA) model that I bought in 1994. It is rather large and bulky, but right after I bought the TV set, I bought a home entertainment center that houses the TV very nicely, along with my receiver, my CD player, and my DVD player. The TV is NTSC-enabled, which means that when the federally mandated switchover to digital actually takes place, my set will be able to receive the digital signals just fine. My TV provides a nice, clear picture, and I have had no problems with it (of course, now I might be jinxing it). Since my TV has long since been paid for, I can continue to use it for free until it dies.

My alternative would be to buy a new TV that is HDTV-enabled, either plasma or LCD. This would cost me roughly $800 for an equivalently sized 32-inch TV. I could mount the new TV on the wall to save space, but then I would have to discard my entertainment center and find a new stand for my receiver, CD player, and DVD player. Other switching costs associated with my buying a new HDTV would include discarding the old TV set, figuring out how to reestablish the rat's nets of cables among the components of my entertainment center, and learning how to use the features of my new HDTV.

11. www.privateline.com/TelephoneHistory2A/Telehistory2A.htm.

According to the HDTV Pub,[12] the benefits I would enjoy from the new TV include

- An aspect (width-to-height) ratio of 16:9 rather than the 4:3 ratio of my current TV. "The increased width is a huge improvement because it allows you to view TV and movies as if you were watching them in a theater. When watching TV on a 16:9 screen you use your peripheral vision, which truly makes you feel like you are a part of the program."
- Up to ten times the resolution and picture quality of my current TV.
- Greater surround-sound capabilities.
- ". . . no more fuzz, no more snow and no more ghosts."
- Progressive, rather than interlaced, scanning, which "provides a sharper picture to the eye and looks much better than an interlaced image of the same resolution."
- Multicasting, which means that television stations will be able to provide more programming over the same channels.

As I mentioned earlier, I tend to be a late adopter of new technologies. Call me a Luddite, but I am very happy with my current TV; for me, the total costs associated with buying a new HDTV are just not worth the added benefits.

More generally, when users replace their old systems with new ones, they effectively pay the *full costs* of the new system ($800 plus switching costs in the case of my HDTV) to accomplish the *extra* things the new system can do that the old system cannot (wider screen and incrementally sharper image and sound system). Users will therefore be willing to adopt the new system only if the *extra value* or *net value* provided by the new system above and beyond the value provided by the old system is greater than the *total costs*, including switching costs, of the new system.

These ideas of net value and net benefits can be expressed using simple visual relationships. Start first with the relationship in Figure 2-3 discussed previously. A more accurate depiction of the user adoption problem in Figure 2-3 would also account for the existence of earlier technology systems and their deterrent effect on the adoption of later systems. I add new terms to Figure 2-3 to capture the existence of currently installed systems to come up with the relationship in Figure 2-8.

---

12. www.hdtvpub.com/articles/whatisdtv/benefits.cfm.

Potential User Chooses to Adopt If
[New Technology Value + New Network Value]
> [Total Cost of New Technology
+ Old Technology Value + Old Network Value]

Figure 2-8

Restated, I end up with a simple relationship in Figure 2-9 that captures the net price and net value ideas. Incremental value in Figure 2-9 is the additional value the new technology system offers over the old system.

> Potential users will adopt a new technology system when the technology plus network values of the new system exceed the technology plus network values of the old system plus new system price plus switching costs.

Recall from the discussion in the previous section that at the beginning of the adoption lifecycle, new technology prices are high and network values are low. During these early stages of technology adoption, the relationships in Figures 2-8 and 2-9 indicate that for new technologies to attract buyers early on, the incremental value the new technology systems offer over the old ones must be high to warrant users giving up their current (old) systems and replacing them with the new systems.

Over time technologies evolve and systems become more complex and more fully featured, and often more durable. It would thus seem to become more and more difficult for suppliers to create new systems whose *adde*d value justifies users replacing

Potential User Chooses to Adopt If
[New Incremental Technology Value
+ New Incremental Network Value]
> [Price of New Technology + Switching Costs]

Figure 2-9

their old systems with the new ones. In actuality, though, studies have shown that adoption lifecycles of more recent technology systems have, in fact, at least kept up with the pace of adoption of older technologies, if not exceeded them.[13] This suggests that system innovators have indeed been able to continue to provide large added values of newer technology systems over older systems.

Suppose that the new system innovator fails to take into account that users currently have a system they are using that performs similar functions to the new proposed technology, though perhaps less effectively or efficiently. In this case, the innovator will erroneously estimate the net new technology value of the new system as the new technology plus network values of the new system, excluding the value the old system is currently providing; that is, the innovator will erroneously omit the **Old Technology Value** and **Old Network Value** terms from the relationship in Figure 2-8. In this case, the new system innovator will overestimate the net value the system brings to users and will thus overestimate the chance of successful adoption of the new system.

> New system innovators who fail to take into account that users currently have systems they are using that perform similar functions to new systems will overestimate their chances of successful adoption of the new systems.

Similarly, a new system innovator might fail to take into account or fully appreciate either (1) the network value provided by the currently established system or (2) the costs of switching from the old system to the new one. In either of these cases, the new innovator will likewise overestimate the value the new system brings into the marketplace and will overestimate the chance of successful adoption of the new system.

> To summarize, there is almost always an existing context or environment into which a new technology is introduced and with which the new technology must compete. Only if the new technology system is *better* at accomplishing the same

13. There have been some claims that technology lifecycles have been decreasing over time; that is, new technologies are being introduced into the market and adopted by users more quickly over time. See, for example, Fraker 1984. Other studies show that faster diffusion of more recent technologies can be explained by other factors, such as availability of infrastructure to support the new innovations (van den Bulte 2000) or increases in product line complexity (Bayus 1998).

thing the old system does, *and at the same time* the price of the new system is worth this *extra* value, will users be led to discard the older system and replace it with the new one. If system innovators fail to take into account the presence of these existing systems, they will overestimate the probability of successful adoption by users of their new systems. Readers interested in further information on technology replacement lifecycles can find a list of resources in Appendix B.

2.4 Critical Mass

After a review of the various lifecycles associated with the introduction of new innovations, it becomes clear that the most essential portion of the process occurs if and when the product begins to take off; that is, the heart of the technology adoption process lies somewhere between acceptance by early adopters and acceptance by the early majority. This is the point during which either the new innovation will become accepted by the masses and take its rightful place in the annals of technology success stories, or the adoption process will fizzle out, and the new innovation will disappear from the market, either never to be heard from again, or to be reinvented and later reintroduced in another form or as part of some other new innovation.

This brings us to the idea of a *threshold point*, *tipping point*, or *point of critical mass*. Once the adoption of a new innovation reaches critical mass, users will increasingly choose to adopt that technology because they expect others to do so, too; that is, further adoption by new users becomes self-reinforcing. Alternatively, for products whose sales fail to reach critical mass, consumers will choose not to adopt that product because they expect others will also not adopt the product. In this case, market share collapses, generally until the product is forced from the market.

Suppose you and a few close friends decide to have a dedicated telephone network (the Hotline) just for this small group; that is, you each have a special phone that will connect only to other members of your small group. If you want to call anyone else, you must use your "regular" telephone. So now what happens? You have to buy a new

> *The tipping point* or point of *critical mass* in the technology adoption lifecycle is that point at which adoption becomes self-reinforcing.

telephone and pay for connection services for the Hotline, and you still have to pay for connection services for your regular telephone. Assuming your phones are cell phones, then whenever you go out, you have to make sure you have both phones with you. After a while, you decide that the Hotline is just too much trouble to maintain. It would be much easier and cheaper to simply have everyone, Hotline friends and regular people, use the same network. So after a few months you decide to disconnect the Hotline. This exemplifies how a very small network generally cannot sustain itself. A few people might initially adopt it, but it's generally not sustainable, in which case it will eventually die out.

At the other extreme, suppose everybody you care about is on the same cell phone network. Suppose then that a few of your contacts leave the network and join a new network, so you can no longer call them on your cell phone. What are your options? You can remain with your current network and be able to call most of your contacts, or you can leave your current network and follow the few contacts who joined the new network. At this point, unless the few people who left are particularly important contacts, you are better off sticking with your current network and being able to access most of your contacts. Furthermore, if everyone else in the old network is in your same position, then no one else in the network will have an incentive to leave the old network and join the new one. Moreover, if someone new comes along, he or she will face the choice of joining the old, large network or joining the new, small network. Unless the few people in the new network are particularly important contacts, the new person will do better by joining the old, larger network. Eventually, the few who left might very well come back to the old network, because the new network might be too small to maintain (see the Hotline example above).

Finally, suppose that the few people who left the large network to start the new network include your boss and your mom. If you cannot reach either of these two people (or they cannot reach you), you will be dead meat. You will then leave the old network and follow Mom and your boss to the new network. Other people in the old network face the same situation as you, so others will also leave the old network and join the new one. In this case, then, what was initially a small network will grow into a large, sustainable network. The old network may survive as a smaller network, or it may end up with too few members and so eventually die out.

This example illustrates how the road to adoption of a new technology will result in one of three possible outcomes: (1) too few people adopt the new technology to support a network that is large enough to be self-sustaining, so the product is forced to leave the market (the new network when Mom and the boss stay with the old network); (2) the network ends up with a few members, enough to support a small network; or (3) the network succeeds in attaining critical mass and ultimately goes on to attract a large number of members (the new network when Mom and the boss join). These three outcomes are illustrated in Figure 2-10.

The two extreme outcomes, those with either no network or a large network, are both *stable*, or *self-reinforcing*. In these two cases, if a few people leave or join the network, the size of the network will tend to be pushed back toward the original point; that is, if a few people break away and start their own network but no one else joins, then the new, small network will not be sustainable, and it will eventually die out. Conversely, if a few people leave a large network, the remaining network will still have plenty of other members to preserve its value, and new users will still find it worthwhile to join.

The second potential outcome, that with a small network, is unstable. If some people leave the network, then the remaining members might not be enough to continue to support the product's use, in which case the product will end up disappearing from the market. If, on the other hand, the network attracts a few additional members, it might trigger the threshold necessary to take off and end up with a widely adopted product with a large network.

Let's return now to two ideas previously discussed. The first idea is that from Figure 2-3: Potential adopters will adopt a new technology if the sum of the technology and network values exceeds the total cost (price plus switching costs). The second idea, that

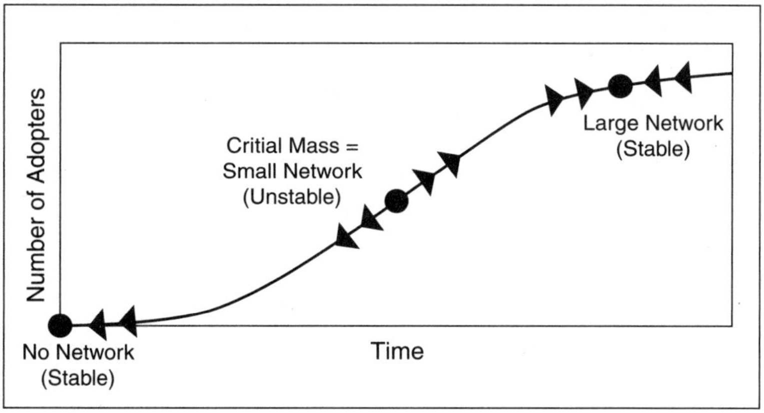

Figure 2-10 Network outcomes

Potential Users Adopt New Technologies If

Technology Value + Network Value > Total Cost of Technology

(Starts High and Decreases) (Starts Low and Increases) (Starts High and Decreases)

Figure 2-11

from Figure 2-7, is that the technology adoption lifecycle is a dynamic process, in which the technology value and the price of the new innovation both fall and the network value rises as the technology moves through its adoption lifecycle. Let's combine these two ideas into a single relationship in Figure 2-11.

Consider the following two-part scenario, a fairly common one, as illustrated in Figure 2-12: (1) Technology values start out high for the early adopters and then drop off quickly with the early majority, while (2) the price starts out high for the early adopters, drops slowly for the early part of the early majority, and then drops faster through a good part of the late majority.

In this case, around the time the tipping point for the adoption of the new technology occurs, the technology's network value must be increasing relatively rapidly for potential adopters to find it worthwhile to buy the new technology; that is, the momentum

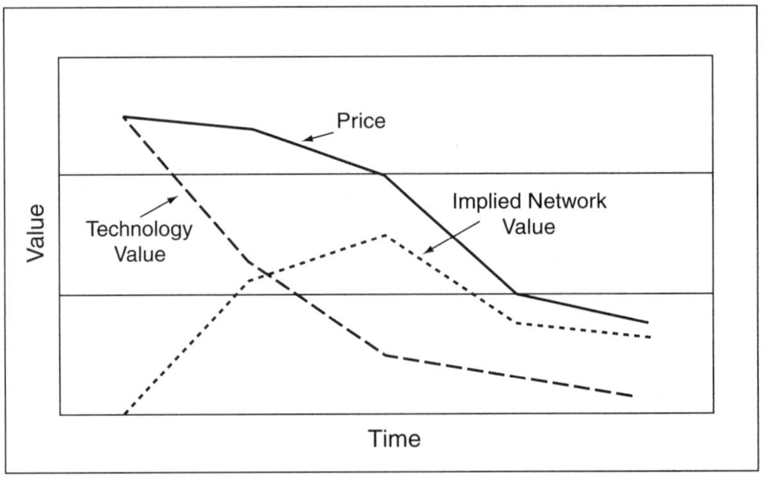

Figure 2-12 Network value dynamics

needed to pull the technology adoption past its tipping point requires that a lot of people join the network (direct network effects are high) and/or that a lot of accessory products and services become available (indirect network effects are high) during this period.[14]

Since technology adoption is based upon how actual and potential users *perceive* the technology, these requirements should really be described as a *perception* or *expectation* that people are joining or will join the network and/or that accessory products and services are becoming available or will become available at some point in the near future. Under an expectations scenario, though, expectations as to what will happen in the future must be strong enough to persuade people to buy the technology now, rather than wait until they see that their expectations do indeed become fulfilled before they buy the new product.

This might come across as being somewhat of a circular argument, that to get people to adopt the new technology, you need to have a lot of people adopting or otherwise creating communities around the new technology. Yet, rather than being a circularity problem, it is actually a close cousin, a chicken-and-egg problem. More specifically, the hardware–software chicken-and-egg problem occurs when users will not adopt a new technology system until there is enough software available, but content providers will not make the software available until there are enough users. What it boils down to is that critical mass will not be achieved until the chicken-and-egg problem has been overcome.

It takes time to develop new accessory products and services for a new technology. Therefore, having the network of accessories available as significant numbers of people are starting to adopt the new technology means that development must actually begin before people have decided to buy the technology; that is, development of accessory products and services needs to start before the central technology has really been accepted by the marketplace. Yet, it is difficult to convince third-party providers to invest in a new technology before the technology has been proven, since it requires a large amount of risk (and faith) on the part of the developers. Most developers would rather decrease the risk by waiting until a greater amount of actual adoption has taken place, and thus the chance that the technology will actually take hold is higher. But

14. This is consistent with Agarwal and Bayus (2002), who note that "the literature finds a sharp takeoff in the number of firms [entering the market for the new technology system] in the early stages of market evolution" (p. 1026).

then you are stuck with having a small network of complementers for the technology when you most need it to be large.

If I avoid this problem by assuming that the momentum required to carry adoption past the tipping point is achieved through direct network effects, rather than indirect network effects, then I still end up with the same problem. People adopt new technologies because others have (or are perceived or expected to have) also adopted them, in which case direct network effects are high. But before people have adopted the new technology, the direct network benefits are small and so is the incentive to adopt.

To summarize, the *threshold point, tipping point,* or *point of critical mass* associated with the adoption of a new technology occurs sometime around the end of the early adopters stage and the beginning of the early majority stage, when further adoption of the technology becomes self-reinforcing; that is, at the tipping point of a product's adoption, users will increasingly choose that product because they expect others to do so. The momentum needed to pull the technology's adoption past its tipping point requires that a lot of people either join or are expected to soon join the network (direct network effects are high) and/or that a lot of accessory products and services become available or are expected to soon become available (indirect network effects are high). Readers interesting in more information on critical mass can find a list of resources in Appendix B.

This leaves us with two paths to the adoption of new technologies:

- The technology itself has large actual or perceived value: The technology value of the new technology is large enough to make adoption worth the price to a significant number of people even when the price is high; that is, stand-alone benefits of the new technology must be large relative to the benefits associated with older technologies.
- The technology has large actual or perceived network effects: In the early stages of adoption, system innovators establish the perception that the technology will soon become widely accepted, so as to generate large network effects (that is, a large number of adopters and/or a large community of resources) early on.

The next section discusses what this means for the Hardware–Software Game.

2.5 Technology Adoption Lifecycles and the Hardware–Software Game

In this chapter I discussed the nature of the five stages of new technology system *production lifecycles*: early stage of production, product takeoff, product competition, product commoditization, and late stage of production. I also discussed the nature of technology system *consumption lifecycles* by four groups of users: innovators and early adopters, the early majority, the late majority, and laggards.

Next, I discussed how the total value of a technology system is equal to the sum of the value of the technology itself, plus the value of the direct and indirect network effects associated with the new technology. Users will choose to adopt the new technology system when the total technology plus network values of the system exceed its price. I also discussed the dynamics of each of these component values over the technology system lifecycle.

I then stressed that for new technologies to attract buyers early on, the incremental value the new technology systems offer over the old ones must be high to warrant users giving up their current (old) systems and replacing them with the new systems.

Finally, I discussed the dynamics surrounding the *threshold point*, *tipping point*, or *point of critical mass* associated with the adoption of a new technology. I established that the momentum needed to pull the technology's adoption past its tipping point requires that a lot of people either join or are expected to soon join the network (direct network effects are high) and/or that a lot of accessory products and services become available or are expected to soon become available (indirect network effects are high).

Let's now return to the Hardware–Software Game and consider how technology adoption lifecycles can help us better understand the incentives of the game's players. Understanding players' incentives will help us predict the game's ultimate results.

The point of critical mass in the adoption of a new technology system generally occurs somewhere between the end of adoption by early adopters and the beginning of adoption by the early majority. Early adopters will not leave their current systems to adopt new ones unless the value of the new system is substantially greater than that of the system currently being used. Since any network value associated with the new system during early stages of adoption is small, this means the new system must offer early adopters some new, distinctive, considerably valuable functionality or some new "killer application" upon purchase that is not available with the older system.

Those in the early majority will not leave their current systems to adopt new ones unless two conditions are met: First, they hear good things about the technology from early adopters, that the new system is meeting expectations; and second, they are assured that supporting infrastructure is well on its way to becoming established, that is, that the innovator has secured commitments from suppliers and/or sponsorship or adoption from large users. Taken together, this means that the new technology system must (1) actually function well and not contain serious bugs or problems, (2) provide users with substantially more value than existing alternatives, and (3) have supporting infrastructure or staying power.

To generate the infrastructure required to draw in the early majority, the hardware manufacturer must provide a sufficient value proposition to attract third-party suppliers of complementary software, services, and accessories; that is, hardware manufacturers must persuade third-party suppliers that the new technology system has significant market potential. In particular, the market potential must be large enough to entice content suppliers away from their established markets into a currently smaller, much riskier market.

These incentive structures are developed more fully in the next two chapters. More specifics as to how system innovators create the incentives for users and third-party suppliers to adopt the new system are discussed in Chapter 6.

CHAPTER THREE

Technology System Users

The first set of players in the Hardware–Software Game that I consider are the technology system users. Technology users must decide if and when to upgrade from the currently installed technology system to the new system. To understand how the game will play out on its own and/or how the speed of adoption of new systems can potentially be increased, we must understand what components affect users' decisions to adopt new technologies, and how these factors change in response to actions taken by hardware and software providers.

Studies on technology adoption and diffusion support the notion that the factors described herein are important drivers of user demand for new technology. More generally, the literature on technology adoption can be divided into three general categories—sociological, marketing, and economic—each of which provides its own perspective on technology adoption. Characteristics of the three approaches are described below.

1. **Sociological Approach:** This body of research views the adoption of new innovations as a social and communicative process, where new innovations are passed from one individual to the next through awareness, education, and validation processes. The research is generally conducted as case studies of the rate of adoption of particular practices or innovations, such as adoption of health programs (sanitation, AIDS prevention, birth control), education programs (teaching methods), agricultural methods (use of fertilizer or hybrid seeds), or new technologies (tools, the Internet). Notable researchers in this area include Thomas W. Valente and Meredith Rolfe.
2. **Marketing Approach:** These studies view the adoption process of new innovations as being driven by word-of-mouth effects and advertising efforts by manufacturers when they launch a new product. In particular, marketing studies seek to emphasize how the characteristics of their products match the needs of

consumers. The research is generally conducted as case studies of the rate of adoption of particular innovations (medical, household, or consumer electronics products). Notable researchers in this area include Frank M. Bass, Barry L. Bayus, Rajesh K. Chandy, Gerard J. Tellus, and Peter N. Golder.

3. **Economic Approach:** Economic studies view the adoption of new innovations as driven by the desires of individuals to maximize their happiness or well-being (utility) and by the desires of companies to maximize profits. This research is generally conducted by focusing on a particular issue that affects the rate of adoption (network externalities, product compatibility, uncertainty); by providing an economic model of the issue, together with the theoretical implications; and perhaps by presenting a specific case study, showing how that experience conforms to the theoretical model presented. Notable researchers in this area include Michael L. Katz, Carl Shapiro, Hal Varian, Stan J. Liebowitz, Stephen E. Margolis, W. Brian Arthur, Nicholas Economides, Joseph Farrell, Garth Saloner, Jeffrey Church, and Neil Gandal.

> Technology users must decide if and when to upgrade from the currently installed technology system to the new system.

I first discuss the factors that influence users' demand for new technology hardware. I then provide a numerical example to illustrate the concepts discussed. Following that I discuss the factors that influence users' demand for new technology software. Then, again, I provide a numerical example for illustrative purposes.

3.1 User Demand for New Technology Hardware

The first mobile telephone service in the United States, American Mobile Phone System (AMPS), was introduced in 1983. As is the case with most new technologies, handsets and connection services were initially expensive and service was spotty. So even though there were not any alternatives to AMPS for people who wanted mobile telephone services, not many people adopted mobile services early on. However, as digital cellular telephone services became more widely available at affordable prices, the number of users skyrocketed, as seen in Figure 3-1.[1] The large demand for cellular services suggests that cellular telephone services provide an appealing alternative for users.

Currently, several alternative mobile services are available, including Verizon (CDMA), AT&T-Cingular (GSM), Sprint-Nextel (CDMA), and T-Mobile (GSM). Assuming all of these providers offer similar features and coverage, each provides a relatively good substitute for the services of the others. This competition in the provision of mobile services constrains the types of services that each of the providers is able to offer (their services must be attractive to buyers) and the prices they can charge (their prices must be low). In this sense, then, although there is a large demand for cellular telephone services generally, the demand for any particular type of cellular service—for example, CDMA or GSM—may not be so large, depending on whether or not it compares favorably on coverage, availability of services and accessories, and price with its close substitutes.

Now let's consider more closely each of the various elements that affect user demand for new technology hardware. A table summarizing the various elements affecting user demand for hardware appears in the summary section at the end of the chapter.

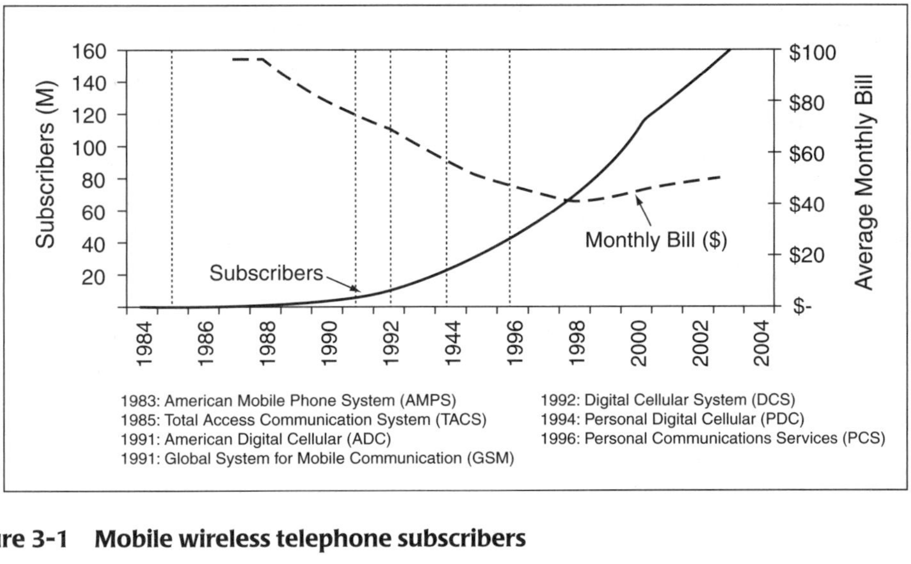

Figure 3-1 Mobile wireless telephone subscribers

1. www.fcc.gov/wcb/iatd/trends.html and www.iec.org/online/tutorials/gsm/topic01.html.

3.1.1 Installed Base

As I have discussed, having more people using a technology system increases its value to users in three ways. First, a larger installed base enables users to share resources or otherwise interact with more users of the system (direct network effects). Second, a larger installed base serves to validate the technology, assuring later potential adopters of its value and encouraging them to adopt. Third, larger installed bases increase value to users by attracting more third-party suppliers of content, software, services, and/or accessories for the technology system, thereby enhancing the value of the system to users (indirect network effects).

The question I now face is this: Suppose someone is currently part of an installed base of users for a particular technology system, and then a new system comes along. To what extent will the size of the installed base of the current system and/or the new system affect that person's decision as to whether (or when) to switch from the current to the new system?

The answer depends on how important each of the three issues just discussed—direct network effects, validation, and indirect network effects—is in creating value for users of each of the two systems.

So although I know *why* users might prefer networks with larger installed bases, I also need to know *how much* the size of the installed base affects user demand. In other words, are the direct network effects associated with a given technology system strong or weak? Does the size of the impact of direct network effects get smaller or larger as the installed base of users grows? That is, I need to know the magnitude or intensity of direct network effects and whether the effects increase or decrease with network size. These two pieces of information will help me better understand why users might prefer one system over another, based on the size of the two systems' respective installed bases.

Consider the first issue, the value of a network's direct network effects. Clearly, a telephone (or fax machine) is no good to someone unless the person he or she wants to call also has a telephone. So the number (and identity) of others with telephones constitutes a significant portion of the value to anyone of having a telephone. A PDA, on the other hand, will provide individual users with plenty of value when they use a PDA technology system by themselves, independent of their ability to link to other users. So, the impact associated with direct network effects on increasing user demand for hardware will be much higher for telephones than it will be for PDAs.

As seen in Figure 3-2, a network with two nodes, Nodes 1 and 2, has two one-way links, one link from Node 1 to Node 2 and one link from Node 2 to Node 1. If I add a third node, Node 3, to the network, then four new one-way links are added (one each to and from Node 3 and Node 1 and one each to and from Node 3 and Node 2), for a total of six one-way links. If we add a fourth node, Node 4, then six new one-way links are added, for a total of 12 one-way links. And if we add a fifth node, Node 5, then eight new one-way links are added, for a total of 20 one-way links.[2] The point here is that when a network is small, new members add few new links, but when networks are large, new members add many new links. And if links create value, then direct network effects associated with larger networks should create more value than those associated with smaller networks.[3]

However, this conclusion requires that *each* member of a network values being linked to *every other* member. In reality, though, most members of a network generally have a relatively well-defined group of people with whom they interact most of the time. Once all the people in that group have joined the network, any new additions to

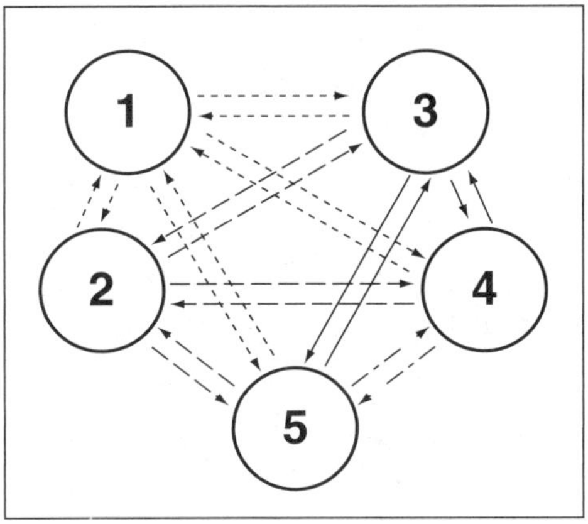

Figure 3-2 Increase in network nodes versus links

2. In general, a network with n nodes contains [n × (n – 1)] one-way links, or [n / (n – 2)] times as many one-way links as a network with (n – 1) nodes.

3. If larger networks provide disproportionately greater value to users than smaller networks, then the sensitivity of user demand to the size of the installed base will be greater than one. If larger networks provide disproportionately less value to users than smaller networks, then the sensitivity of user demand to the size of the installed base will be less than one.

the network will create little, if any, value for them.[4] If this does, in fact, provide a better description of the real world, then when networks are small, each new member will provide significant value to existing members. Yet, over time, as the networks expand and new members continue to join, the increase in value to existing members will be relatively small.[5]

> Having a larger installed base will increase user demand for hardware more during the earlier stages of adoption, and the impact will decrease as more users adopt.

The second way the size of an installed base might affect user demand for technology systems is by validating the technology, that is, by assuring later adopters of its value and encouraging them to adopt. This validation impact on user demand will be technology-specific, and it will be more important for technology systems that are more complex and/or whose adoption by users is riskier. *Riskier systems* are those systems that are more expensive and/or have greater probabilities of not meeting expectations, with the result of potentially leaving users stuck with systems that provide less value than the total costs (time, money, and other costs) they paid.

Frank Bass, one of the pioneers of technology adoption models, refers to early adopters as *innovators* and later adopters as *imitators*, who "'learn' in some sense, from those who have already bought. The importance of innovators will be greater at first but will diminish monotonically with time."[6] In other words, as more and more people adopt a system, those who have yet to adopt become more and more convinced of the system's value. However, at some point, the number of adopters reaches a size at which

4. Metcalfe's Law asserts that the value of a network with n nodes grows proportionately with n^2, which is akin to saying that every link between every node in a network has value. The problem with this assertion is that it implies that only one network will survive, since one larger network would necessarily provide greater value than two smaller networks. Yet, in reality, we see industries in which multiple networks coexist. Odlyzko and Tilly (2005) propose that the value of a network with n nodes grows in proportion to (n log n), which says that people tend to value only a subset of people on the network, and not everyone. This relationship allows for the coexistence of multiple networks in an industry.

5. In other words, the exponent on the size of the installed base will be between zero and one: The impact of the size of the installed base on increasing demand for new hardware will increase with the size of the installed base, but at a decreasing rate.

6. Bass 1969.

the system has been fully validated. After this threshold point has been reached, further adoption by more users does not help to further convince those who have not yet adopted the system that the system is valuable.

The last way larger installed bases might affect user demand is indirectly, by attracting more third-party suppliers of system content, thereby enhancing the value of the system to users. In my analysis, I account for the impact of content availability on user demand for hardware by *directly* accounting for content availability, which is further discussed below.

Let's return to the case of mobile telephone services. As of mid-2007, there was a total of 243.4 million subscribers for all cellular service providers in the United States, representing an 81 percent overall penetration rate,[7] with market shares of roughly 60 percent CDMA and 40 percent GSM.[8]

Suppose a user had signed up for second-generation cellular service early on with T-Mobile and GSM. The contract is now up and the user has to decide whether to renew it on the GSM network or switch over to CDMA. Would the fact that the CDMA network is larger than the GSM network lead the GSM user to switch to the CDMA network?

When it comes to cellular service, the big consideration that tends to affect users' decisions about which systems and/or providers to use is network coverage (Can I make and receives calls from where I am?). In addition to coverage, pricing (What types of pricing packages do they offer? Are there roaming fees?) and available features and accessories (What is the battery life of my phone? Can I access the Internet? Can I send and receive text messages? Can I take pictures?) are also important.

The answer to the question is that if callers on the GSM network cannot get the same accessibility to service in *their area* by using a CDMA system, then they most likely would not switch networks, even though the CDMA system might offer more coverage overall. However, if GSM and CDMA offer the same accessibility in their area, then they might switch from GSM to CDMA, if CDMA, being the larger network, offers greater access, more features, and/or lower prices than the GSM system, net of switching costs (the costs of buying a new CDMA handset).

---

7. www.ctia.org/media/industry_info/index.cfm/AID/10323.

8. http://en.wikipedia.org/wiki/List_of_mobile_network_operators_of_the_Americas#United_States\. TMDA and other systems are provided by some operators who also provide service for CDMA and/or GSM systems. For simplicity's sake, I ignore these other systems.

To summarize this section, the three means by which the size of a technology system's user base might increase hardware demand by new adopters are providing direct network effects, validation, and indirect network effects. The first two routes should have a larger impact on increasing demand early on when there are few members of the network; they should have a smaller impact on increasing demand later when there are more members of the network. As for the third route, indirect network effects, I capture those effects elsewhere. So based on the two routes left, I should generally expect the size of the installed base to have a larger impact on increasing new adoption of the system hardware early on. As the size of the user base grows, the size of the installed base should have a diminishing impact on increasing new adoption.

There is a further note on the dynamics of installed bases. When there are direct network effects, as more users and content switch over from an old technology system to a new one, the size of the installed base for the older system shrinks. The old technology system correspondingly becomes less valuable to users who have yet to switch. This dynamic will tend to create momentum, so that once the size of the installed base for a new system reaches critical mass, direct network effects will simultaneously attract new users to the newer system and repel old users from the older system.

3.1.2 The Stock of Content Available

In the technology systems I am considering, the availability of more or different kinds of content (software and accessories) enhances the user experience associated with a technology system. In this case, user demand for the hardware will be greater when there is more content available.

Analogous to the situation with the installed base, the question I face is this: Suppose someone is currently part of an installed base of users for a particular technology system, and then a new system comes along. To what extent will the availability of content for the current and/or new system affect that person's decision as to whether (or when) to switch from the current to the new system?

As with the installed base, the answer depends on how important the availability of content is for enhancing the value of the user experience associated with the technology. In the technology systems I am considering, content increases the value of the system to users. By definition, then, the demand for new technology hardware is greater when there is more content available. But how much greater? Does having twice as much content double the demand, or does it increase demand by, say, only 10 percent? And is the impact the same no matter how much content is available, or does it change with the amount of available content?

Return once more to the evolution of cellular telephony, as displayed in Figure 3-1,[9] and consider how demand for content has changed as telephony has evolved. First-generation mobile systems (NMT, AMPS, TACS, and RTMI) became available during the early 1980s. The Motorola DynaTAC 8000X was the world's first commercial hand-held cellular phone, with a price tag of $3,995 in 1983 dollars, or about $8,400 in 2007 dollars. The phone was large and bulky, enabled 30 minutes of talking time, and took ten hours to recharge. Transmissions were in analog, and available content and accessories essentially included coverage of service, which was patchy across the nation, not to mention very expensive. Given the high total cost of using the first-generation system, there were relatively few users, and demand for hardware did not change much with system coverage.

Second-generation cellular systems (GSM, TDMA, and CDMA) became available during the early 1990s. Transmissions were digital, cell phones became much smaller in size, and content offerings expanded to include (1) service coverage; (2) different models of cell phones, each with their respective features; (3) cell phone accessories, including extra-life batteries and carrying cases; and (4) more varieties of services, including different pricing packages of monthly minutes and different features, such as voice mail, call forwarding, and conferencing calling. With the greater variety of available services and accessories and the lower prices of handsets and service, users of second-generation cell phones were given many more choices. Accordingly, demand for any particular variety of cell phone handset became more sensitive. It now mattered what variety of services and features (jointly referred to as *content*) was available for that handset, as compared with the content available for other handsets.

Third-generation cellular systems (UMTS and CDMA2000 1xEV-DO) are now becoming available. In addition to all the content and accessories for second-generation cellular telephony, third-generation services will also include live streaming radio and

9. www.fcc.gov/wcb/iatd/trends.html, www.iec.org/online/tutorials/gsm/topic01.html, www.retrobrick.com/moto8000.html, http://en.wikipedia.org/wiki/History_of_mobile_phones.

video services. Sensitivity of demand for third-generation cell phones should correspondingly increase over that for second-generation cell phones.

As long as users value content, the demand for hardware will be greater when there is more content available. However, the size of the impact may be larger or smaller for different technologies, and the size of the impact may increase or decrease over time with the amount of content available, depending upon whether later content complements or substitutes for earlier content. (This last concept is discussed in more detail at the end of section 3.2.2.)

Before Palm introduced the Palm Pilot, I used to carry around a day planner, so I could keep track of all my contacts and appointments. When the Palm Pilot came out, I was so ready to exchange my messy day planner for the sleek, simple Palm Pilot. As the Palm became more popular, other applications became available, such as games and spreadsheet programs. However, I am not a big fan of computer games, and I would rather do my spreadsheets and word processing on my computer, so I never bought any of the other applications available for the Palm Pilot.

Users buy most technology systems for specific applications or general areas of application. I bought my Palm Pilot to keep track of contacts and appointments. If there are enough other people out there like me, then the introduction of different applications for the Palm, such as computer games and spreadsheets, will not cause a whole lot of other people to go out and buy Palm Pilots. People buy hammers to hammer in nails, they buy stereo systems to listen to music, and they buy copy machines to make photocopies. In these cases, having more content available for the hardware (having nails in different sizes, music in different genres, paper and ink in different colors, or different applications for the Palm Pilot) will lead some more people to buy the hardware. However, the increase in new hardware sales brought about by the introduction of new content will generally be relatively small.

In other cases, though, having more content available for technology systems can lead to much larger impacts on the demand for hardware. In particular, when completely new applications are found for technologies, the availability of content for new applications will lead whole new groups of users to buy the hardware. This is what happened with the earlier adoption of CD players by popular and rock music listeners, and then the later adoption of CD players as data storage media:

> *As early as 1979, Philips and Sony set up a joint task force of engineers to design the new digital audio disc . . . Piet Kramer, who at the time was a member of the optical group at Philips that made a significant contribution to the CD technology, commented . . . :*

> *". . . we never imagined that one day the computing and entertainment industries would also opt for the digital CD for storing the growing volume of data for computer programs and movies."*[10]

The same multi-application phenomenon also unfolded in the market for Apple iPods. Early adopters initially bought iPods to listen to music, but as more people have adopted iPods, new and different content has become available. As of November 12, 2007, the Podcast Directory[11] contained 18 general categories, including News & Media, Home & Lifestyle, Politics & Government, and Soliloquies, containing more than 39,617 podcasts. As such, there must certainly be plenty of iPod users who rarely use their iPods to listen to music and instead listen mostly just to podcasts.

> To summarize, some technology systems are more dependent than others on having a large variety of software or content to make the technology system valuable. As a result, the sensitivity of user demand for hardware to the availability of content will depend on the particular system at issue. Also, the impact of content availability on the demand for hardware may increase or decrease as more content becomes available, depending on whether you can stand only so much (content saturation), or you can never get enough of a good thing (new applications or new user groups).

3.1.3 Price of New Technology Hardware and Content

Suppose you have a vacation coming up and you want to buy a Nintendo Game Boy so you can play the Final Fantasy I–V games during your break. The price for the handheld device is $90 and the price for the game cartridges is $20 each, for a total price of $190. Would you be happier or more willing to buy the system if the price of the handheld were $40 and the price of the game cartridges were $150? Probably not. What you care about is the total price of playing Final Fantasy, that you have to pay your $190 up front, and that you will be able to immediately enjoy the games. Beyond that, it really does not matter much how that total price is distributed between the hardware and software components.

10. http://origin.newscenter.philips.com/about/news/press/20070816_25th_anniversary_cd.page.

11. www.podcast.net/.

Alternatively, suppose you buy a Nintendo Game Boy so you can play the Final Fantasy I–V games, but now you plan on buying the gaming cartridges one by one as you get tired of each of the individual serials. The price for the handheld device is $90, and the price of the first cartridge is $20, for an initial price of $110, to be followed up by subsequent payments of $20 for each of the remaining four cartridges. The total price for the handheld device and the five game cartridges is still $190, but now you are paying $110 up front, with the option of paying more later for the other games. Are you still indifferent to how the total $190 price is allocated between the handheld and the game cartridges? Probably not. Now you would prefer that the price of the handheld device be lower and the price of the gaming cartridges be higher. The difference has to do with the timing of when you must pay for the technology system relative to when you enjoy its benefits. In this case, you would rather pay $40 for the handheld and $30 for the first cartridge, for a total of $70 up front. Then, if you later decide the system is not as much fun as you thought it would be, and so you do not buy the other cartridges, you have paid less for the experience than you would have with the initial price allocation ($70 versus $110).

More generally, users care about the risk associated with buying a new technology. The risk is that the cost of the system will end up being greater than the value users eventually get from it. The costs associated with buying and installing a new technology system will be riskier when (1) the up-front costs associated with buying and installing a technology system (including any switching costs) are larger—with larger costs, the benefits must be larger so as to make the purchase of the technology worthwhile; (2) the costs occur earlier relative to the time in which the consumption experience is generated—because of the time value of money, costs incurred later are discounted relative to costs incurred earlier; or (3) the system has not yet been proven to be effective; that is, not enough people have adopted the system to provide sufficient validation. The implications are that users will tend to be more sensitive to the price of hardware than to the price of content (assuming they buy hardware before they buy content), and their sensitivity will decrease over time as the risks associated with using the system decrease.

> Users will tend to be more sensitive to the price of hardware than they will be to the price of content, and their sensitivity will decrease over time as the risks associated with using the system decrease.

Similarly, before users commit to any particular system, they tend to have a lot of choices—all the components across the various systems—in which case they will be more sensitive to the prices of the individual system components. However, after users have committed to a particular system by purchasing the hardware, they no longer have the same software/content/accessory choices that they had before they decided on that particular system. Rather, they are limited to only that content that is compatible with the particular system they chose. This reinforces the notion that user sensitivity to the price of hardware and content components tends to be higher before users buy a new technology system than after they buy a new system. If users tend to buy hardware before they buy software or content, then users will tend to be more sensitive to the price of hardware and software components when they buy the hardware than when they buy content.

Another big issue has to do with the new technology's *compatibility,* in the social sense, as described by Everett Rogers,[12] as well as in the technical sense. According to Rogers, a large part of the appeal of a new technology is how well it accords, or is *compatible*, with users' beliefs, experiences, and lifestyles. Users have become increasingly interconnected with one another within all types of networks, be they professional, social, political, and so forth. As such, maintaining network connectivity has become a larger consideration when deciding whether or not to adopt a new technology. This is just another way of saying that the network component of the total value of new technology has been increasing over time. At this point, the network component has become a significant concern for users when deciding if and when to adopt a new technology system. New technologies that are not compatible with existing systems—that do not easily interlink to current systems—will be more costly for users to adopt; that is, switching costs will be higher.

The discussion on social compatibility and switching costs has an interesting implication. The desire to stay connected with one's network will tend to decrease the speed of adoption during the early stages of the adoption lifecycle, and it will tend to increase the speed of adoption during the later stages; that is, network technologies will initially tend to start gathering users later than non-network technology systems. However, once adoption gains momentum, network systems will end up experiencing faster adoption by the majority than non-network systems.

12. Rogers 2003.

> As compared with non-network systems, network technologies will tend to attract new adopters more slowly initially, but more quickly later on.

I previously indicated that early adopters tend to be less sensitive to price than later adopters. As users move through the adoption cycle of a new technology system, from early adopters through the early majority, the later majority, and finally laggards, one would thus expect users' sensitivity to price to *increase* over time.

However, I also discussed how the risks and uncertainties associated with new technology systems make users sensitive to system prices, where higher prices mean more risk. As users move through the adoption lifecycle of a new technology system, from first-time users through validation and repeat purchases, one would expect that as the risk associated with new adoption dissipates, users' sensitivity to price would *decrease* over time.

Empirical analyses show that over time there are more repeat purchases of less durable products—that is, products with shorter life spans—than new purchases. In these cases, validation effects outweigh the early-late adopter effects, so price sensitivities tend to decrease over time. For more durable products, though, over time, first-time purchases tend to dominate repeat purchases even for long time periods, so the early-late adopter effects outweigh validation effects, and price sensitivities tend to increase over time.[13]

Taken together, these two phenomena of decreasing user sensitivity to component prices before versus after committing to a system, and increasing user sensitivities to price over the adoption lifecycle, suggest that users will be more sensitive to the price of hardware than of software, and that price sensitivities for both hardware and software will increase over time for more durable systems and decrease over time for less durable systems.

> Users' price sensitivities for both hardware and software will increase over time for more durable systems and decrease over time for less durable systems.

13. See, for example, Danaher, Hardie, and Putsis 2001; Parker and Neelamegham 1997; and Parker 1992.

To summarize, users will tend to be more sensitive to the price of hardware than of software. First-time buyers will tend to be more sensitive to price as adoption progresses from early adopters, through the early and late majorities, to laggards. Repeat purchasers will tend to be less sensitive to all prices over time as the system becomes validated. As networks become more established and switching costs increase, network technologies will be slower to begin adoption than non-network technology systems. Yet, once adoption gathers momentum, the majority will adopt network systems faster than non-network systems.

3.1.4 Numerical Example of Demand for New Technology Hardware

Recall the relationship in Figure 2-8, which says that potential users will adopt a new technology system if and when the value of the new technology and its associated networks, net of costs, exceeds the value of the existing technology plus its associated networks. Additionally, in the previous section I described conceptually how the value of the new technology and its associated networks net of costs, relative to the value of the old technology and its associated networks, affects user demand for the new technology hardware. I also stressed that if technology innovators fail to account for the value provided by currently installed systems, they will tend to overestimate the value their new systems provide, which will lead them to overestimate the expected speed of adoption of their systems. I now provide some numerical examples to better illustrate these concepts.

Based on the relationship in Figure 2-8, I have come up with a simulation model[14] that will enable me to use actual numbers to see what a real-word user demand function might look like. In the first example, which I call the Old & New Technology (O&NT) Case, I consider the case of a new technology system that is introduced into a market in which another system is already being used. In this O&NT Case the technology system innovator correctly accounts for the existence of an installed system. Using the simulation model, I show how changes in the size of the installed base and in the availability of content affect user demand for hardware.

14. The model is presented in more detail in Appendix A. I apply the user demand function for hardware, equation (C-1), in the two examples discussed here, the O&NT Case and the NT Case.

I then look at another example, the New Technology (NT) Case, in which a technology system innovator fails to account for the existence of an older installed technology system that already provides services similar to those that the new system offers. Comparing the results of the NT Case with those found in the O&NT Case, I show that in the NT Case technology innovators would predict that their technology would be adopted sooner than if they were to account for the installed system.

3.1.4.1 Assumptions

I assume there are 1,000,000 people in a particular area, and I assume that the total amount of content available each period to be divided among technology systems is 100 units (for example, 100 games, 100 programs, 100 channels, and so on).

Conceptually, I assume the following:

- Newer technology systems, with more features and benefits, appeal to more potential users than older systems.
- The impact of network effects decreases over time, so that newer systems and smaller networks will exhibit greater network effects.
- Newer systems are riskier than older, more established ones.
- Newer systems are more expensive than older ones.

Numerically, I assume the following for the *currently available technology system:*

- The potential market size is 100,000; that is, under the right circumstances, 100,000 people in the population would potentially adopt the current technology.
- The sensitivity of hardware demand to the size of the installed base is 0.40; that is, if the size of the installed base increases by 10 percent, then demand for hardware by new adopters will increase by 4 percent.
- The sensitivity of hardware demand to the amount of content available is 0.40; that is, if the supply of content for the new system increases by 10 percent, then demand for hardware by new adopters will increase by 4 percent.
- The sensitivity of hardware demand to the price of hardware is –2.75; that is, if the price is 10 percent higher, then demand for hardware by new adopters will decrease by 27.5 percent.
- The price of current system content is $10 per unit.

And I assume for the ***new technology system*** that is subsequently introduced into the market:

- The potential market size is 1,000,000; that is, under the right circumstances, 1,000,000 people in the population would potentially adopt the current technology.
- The sensitivity of hardware demand to the size of the installed base is 1.20; that is, if the size of the installed base increases by 10 percent, then demand for hardware by new adopters will increase by 12 percent.
- The sensitivity of hardware demand to the amount of content available is 1.20; that is, if the supply of content for the new system increases by 10 percent, then demand for hardware by new adopters will increase by 12 percent.
- The sensitivity of hardware demand to the price of hardware is –3.50; that is, if the price is 10 percent higher, then demand for hardware by new adopters will decrease by 35 percent.
- The price of new system content is $20 per unit, and the price of new system hardware is $300.

3.1.4.2 Old & New Technology Case

In the Old & New Technology (O&NT) Case the new system innovator takes into account the fact that there are other systems currently available for doing similar things to the new system. In this case, the system innovator recognizes that users will adopt the new system only if its *added* value over the current system is worth the price, as seen in the relationship in Figure 3-3.

Users Will Adopt New Technology System If

{ Technology Value of New System + Value of Installed Base for New System + Value of Software for New System }

> { Technology Value of Old System + Value of Installed Base for Old System + Value of Software for Old System }

\+ { Price of New System + Switching Cost }

Figure 3-3

I plug the numbers indicated in my assumptions into the model. The model outputs the relationships displayed in Figures 3-4 and 3-5 as the expected portion of potential adopters who will adopt the new technology for different levels of previous adopters and content availability. For example, my model says that if 10 percent of content is provided for the new technology system (and 90 percent is provided for the old system) and if 40 percent of potential adopters of the new system have already adopted it, then at a price of $300 for the new hardware, no new potential adopters will buy the new hardware. If the installed base is larger, so that 60 percent of potential adopters of the new system have already adopted it, then the value of the new system

Portion of New Adopters											
Portion of Users with New Technology	Portion of Content Supplied for New Technology										
	1%	10%	20%	30%	40%	50%	60%	70%	80%	90%	99%
1%	0.00	0.00	0.00	0.00	0.00	0.00	0.00	0.00	0.00	0.00	0.00
10%	0.00	0.00	0.00	0.00	0.00	0.00	0.04	0.11	0.19	0.27	0.38
20%	0.00	0.00	0.00	0.01	0.12	0.23	0.35	0.48	0.62	0.76	0.93
30%	0.00	0.00	0.01	0.17	0.33	0.51	0.70	0.89	1.00	1.00	1.00
40%	0.00	0.00	0.12	0.33	0.57	0.81	1.00	1.00	1.00	1.00	1.00
50%	0.00	0.00	0.23	0.51	0.81	1.00	1.00	1.00	1.00	1.00	1.00
60%	0.00	0.04	0.35	0.70	1.00	1.00	1.00	1.00	1.00	1.00	1.00
70%	0.00	0.11	0.48	0.89	1.00	1.00	1.00	1.00	1.00	1.00	1.00
80%	0.00	0.19	0.62	1.00	1.00	1.00	1.00	1.00	1.00	1.00	1.00
90%	0.00	0.27	0.76	1.00	1.00	1.00	1.00	1.00	1.00	1.00	1.00
99%	0.00	0.38	0.93	1.00	1.00	1.00	1.00	1.00	1.00	1.00	1.00

Figure 3-4 User demand for hardware example: table of portion of new adopters (O&NT Case)

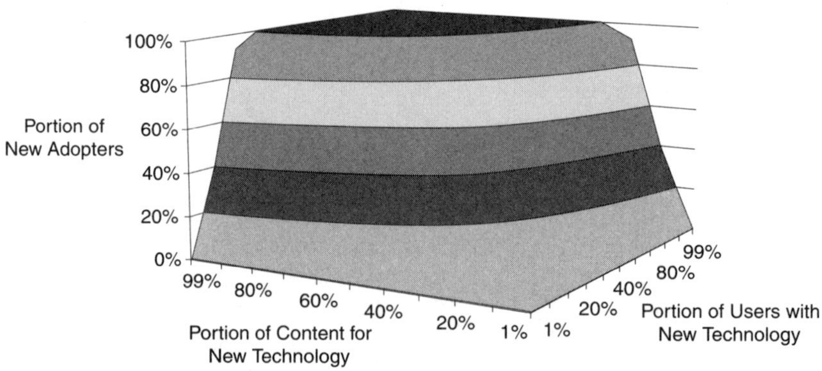

Figure 3-5 User demand for hardware example: graph of portion of new adopters (O&NT Case)

increases due to direct networks effects, and now 4 percent of potential users will adopt the new system at a price of \$300.

This example illustrates the two basic relationships I have discussed. First, moving across the columns of the table, when there is more content, the value of the technology system to users increases. This means that with more content available, new adopters are willing to pay a higher price for new technology hardware, or equivalently, the hardware manufacturer can sell more hardware at any given price. And second, moving down the rows of the table, when there are more users for the new system, the value of the technology system to users is greater. With more users, then, new adopters are willing to pay a higher price for new technology hardware, or equivalently, the hardware manufacturer can sell more hardware at any given price.

3.1.4.3 New Technology Case

In the New Technology (NT) Case the new system innovator looks at the adoption of the new system in isolation; that is, the innovator fails to take into account the fact that there is another system currently available for doing similar things to the new system. In this case, the system innovator erroneously believes that users will adopt the new system if its total value, rather than just its added value over the alternative system, is worth the price, as seen in the relationship in Figure 3-6.

I plug the numbers mentioned earlier into the model, and I get the relationships displayed in Figures 3-7 and 3-8 for the expected portion of potential adopters who will adopt the new technology for different levels of previous adopters and content availability.

Comparing the results for the NT Case (technology adoption in isolation) and the O&NT Case (technology adoption considering available alternatives), the model shows that users will adopt the new technology faster in the NT Case. More specifically, if 10 percent of content is made available for the new technology and 40 percent of potential

Adopt New Technology System If

$$\left\{ \text{Technology Value of New System} + \text{Value of Installed Base for New System} + \text{Value of Software for New System} \right\} > \text{Price of New System}$$

Figure 3-6

Portion of New Adopters											
	Portion of Content Supplied for New Technology										
Portion of Users with New Technology	1%	10%	20%	30%	40%	50%	60%	70%	80%	90%	99%
1%	0.00	0.00	0.00	0.01	0.01	0.01	0.01	0.02	0.02	0.02	0.03
10%	0.00	0.03	0.06	0.10	0.14	0.19	0.23	0.28	0.33	0.38	0.42
20%	0.00	0.06	0.14	0.23	0.33	0.43	0.53	0.64	0.75	0.87	0.97
30%	0.01	0.10	0.23	0.38	0.53	0.70	0.87	1.00	1.00	1.00	1.00
40%	0.01	0.14	0.33	0.53	0.75	0.98	1.00	1.00	1.00	1.00	1.00
50%	0.01	0.19	0.43	0.70	0.98	1.00	1.00	1.00	1.00	1.00	1.00
60%	0.01	0.23	0.53	0.87	1.00	1.00	1.00	1.00	1.00	1.00	1.00
70%	0.02	0.28	0.64	1.00	1.00	1.00	1.00	1.00	1.00	1.00	1.00
80%	0.02	0.33	0.75	1.00	1.00	1.00	1.00	1.00	1.00	1.00	1.00
90%	0.02	0.38	0.87	1.00	1.00	1.00	1.00	1.00	1.00	1.00	1.00
99%	0.03	0.42	0.97	1.00	1.00	1.00	1.00	1.00	1.00	1.00	1.00

Figure 3-7 User demand for hardware example: table of portion of new adopters (NT Case)

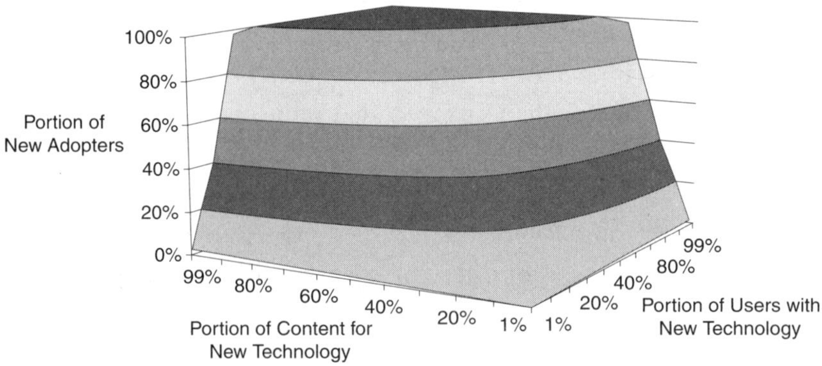

Figure 3-8 User demand for hardware example: graph of portion of new adopters (NT Case)

users have purchased the new technology hardware, then the NT Case predicts that 14 percent of potential users who have not yet adopted the new technology system will now choose to adopt, as compared with no users in the O&NT Case. Taken together, these two examples illustrate how system innovators will overestimate the speed of adoption of their new systems if they fail to account for the fact that users have other ways of doing what the new technology does.

3.2 User Demand for New Technology Content

I have just discussed how user demand for new technology hardware works. Now I take on user demand for new technology software or content. Demand for content works in much the same way as demand for hardware, except that demand for software is less complicated. Since demand for new system content comes only from actual users of the new system, it is not affected by the existence of other systems, as is the demand for new hardware.

In the case of hardware, users buy the new technology hardware when the net value provided by the system exceeds its total cost (hardware price plus switching costs). Similarly, in the case of software or content, users buy new technology content when the value it provides exceeds its cost. How does content provide value? A user will combine content with hardware (hardware owned by the user and perhaps hardware owned by other linked users), and possibly other content (content owned by the user and perhaps content owned by other linked users), to achieve the consumption experience provided by the system. This means that the value provided by new system content will depend on the amount of hardware in use, the amount of software in use, and the price of software. The remainder of this section discusses the impact on user demand for content of increases in the size of the installed base and in content availability. The discussion for the price of content is included above with that for the price of hardware. A table summarizing these three elements affecting user demand for content appears in the summary section at the end of the chapter.

3.2.1 Installed Base

I have a calculator that I use to balance my checkbook. It's nothing fancy; it has all the basic functions, and not much else, but what more would you do with a calculator? I suppose I could buy a printer for it, but I do not really need a printer for my calculator. For similar systems in which content does not play a large role in creating value for users, demand for content will increase only slowly with the size of the installed base. Other examples of systems whose value is primarily based on hardware include kitchen appliances and hand tools. Accessory attachments may be available for these technology systems, but most users are generally satisfied with only electricity or batteries to power the hardware.

In contrast, my iPod needs a lot of content for it to be useful. It doesn't do me any good unless I have music or podcasts to play on it. As such, when content plays a large

> The impact of a larger installed base on increasing user demand for content will tend to decrease over time as users become saturated with content.

role in creating value for system users, demand for content can increase quite dramatically with the size of the installed base of users. This will be the case when the technology hardware serves more as a conduit for accessing software than as a stand-alone device. Examples of technologies for which hardware serves as a means of accessing content include CD and DVD players, TVs, cell phones, and gaming devices such as Sega, Nintendo, or PlayStation. Content will also be more important in creating value for users when it plays a large role in customizing relatively standardized hardware for relatively heterogeneous users. Examples include transportation networks, where different users access different roads, railways, and air routes; and ATM machines, where different users access the machines in different locations.

> To summarize, for technology systems in which content does not play a large role in creating value for users, demand for content will increase only slowly with the size of the installed base. In contrast, for technology systems in which content does play a large role in creating value for users, demand for content will increase quickly with the size of the installed base. Regardless of whether content creates little or much value for system users, however, the impact will tend to decrease over time as users become saturated with content.

3.2.2 The Stock of Content Available

One of my favorite technology systems from a timesaving perspective is the Universal Product Code (UPC) stamp and bar-code scanner. The first item sold using a bar-code scanner was a pack of Wrigley's gum at a Marsh supermarket in Troy, Ohio, on June 26, 1974. The early bar-code systems were first adopted by grocery stores and used simply to automate the checkout process, for which it was estimated that the bar-code system enabled cashiers to check out items twice as fast as they could using traditional equipment. Only later, after most items came labeled with bar codes, were supermarkets able to use bar codes for tracking inventory.[15]

Today just the basic applications for bar-code scanners enable grocery stores to manage inventory (receiving, shipping, tracking), check out items, and change prices automatically. In 2007 the average number of items carried in a typical supermarket was about 45,000, and the average number of inventory turns per year was about 14.[16] So the average supermarket now sells roughly 630,000 (= 45,000 × 14) items per year. For every second of time per item a bar-code scanner saves a grocery store in pricing, checking out, receiving, shipping, inventorying items, and so on, each year, that grocery store will save 175 hours, or 7.3 days.[17] That's a lot of savings!

In the case of bar-code scanners, applications generally tend to complement each other; that is, once a scanner system is installed, different applications enable grocery stores to accomplish a wide variety of different tasks, including processing credit cards, managing inventory, and analyzing point-of-sale data. As new applications become available, grocery stores will probably choose to avail themselves of many of them since they enhance the value of owning and using a bar-code scanner system. In other words, user demand for bar-code system content will tend to increase relatively quickly with available supply.

Alternatively, consider the case of technology systems for which content consists mainly of consumables, such as copy/fax machines and paper/toner, hammers and nails, paintbrushes and paint, automobiles and gas/maintenance. In these cases, having some amount of content (standard white paper, black ink, 3/8-inch nails, white latex paint, regular unleaded gasoline) provides tremendous value, but then having more content (heavy bond paper, red ink, brads, pink high-gloss paint, high-octane gasoline) does not add much more value. In these cases, the added content simply substitutes for the earlier content. For these types of technologies, where having some content is good, but having even more content is just somewhat better, demand for content will increase only slowly with the available supply.

---

15. Seideman 1993.

16. Food Marketing Institute, www.fmi.org/facts_figs/faq/?fuseaction=faq_18 and www.fmi.org/facts_figs/keyfacts/?fuseaction=inventory.

17. 630,000 items per year × 1 second per item × 1 minute / 60 seconds × 1 hour / 60 minutes = 175 hours.

> Having more content available will increase user demand for content more or less over time as later content complements or substitutes for earlier content.

More generally, the degree to which the supply of available content for a technology system affects its demand by users depends on the degree to which having more content provides more value for the technology system. For systems from which value achieved by users is substantially higher when there is a large supply or a large variety of content available, or if different units of content tend to complement one another, demand for content will increase more rapidly with supply. If, on the other hand, the technology system is not very dependent on content for creating value, or if different units of content tend to substitute for, rather than complement, one another, then user demand for content will increase only slowly with the available supply.

3.2.3 Numerical Example of Demand for New Technology Content

The preceding sections describe conceptually the three factors that affect user demand for new technology: the amount of hardware in use, the amount of software in use, and the price of software (discussed in section 3.1.3). I now present a numerical example to help illustrate these concepts. Continuing with the model introduced in section 3.1.4, let's plug in some numbers to get a picture of how user demand for content works.[18]

Using the same assumptions as before, there are 1,000,000 people in a particular area, and the total amount of content available each period is 100 units (for example, 100 games, 100 programs, 100 channels, and so on) to be divided among technology systems.

Numerically, I assume the following for the ***new technology system:***

- The potential market size is 1,000,000; that is, under the right circumstances, 1,000,000 people in the population would potentially adopt the current technology.

18. The model is presented in more detail in Appendix A. I apply the user demand function for content, equation (C-2), in the example here.

- The sensitivity of software demand to the size of the installed base is 0.85; that is, if the size of the installed base increases by 10 percent, then demand for software will increase by 8.5 percent.
- The sensitivity of software demand to the amount of software available is 1.20; that is, if the supply of content for the new system increases by 10 percent, then demand for software will increase by 12 percent.
- The sensitivity of software demand to the price of software is –2.75; that is, if the price is 10 percent higher, then demand for hardware by new adopters will decrease by 27.5 percent.
- The price of content is $25 per unit.

I plug these numbers into the model, and I get the relationships displayed in Figures 3-9 and 3-10 for the prediction of the portion of content that will be demanded for the new technology for different levels of previous adopters and content availability.

Figure 3-9 says that if 40 percent of users have upgraded to the new technology system, and if 10 percent of content was offered for the new system last period, then this period demand for new system content will be 7 percent. If the same amount of content was offered for the new system last period (10 percent), but more users have upgraded to the new system, say, 60 percent, then demand for new system content this period will be greater, 9 percent, rather than 7 percent. More generally, for the upper left diagonal portion of the table, where there is relatively more content than there are users, the content provider should decrease the supply of content for the new system and/or decrease the price. The opposite is true for the lower right diagonal portion of the table; that is, where there are relatively more users than there is content, the content provider should increase the supply of content for the new system and/or increase the price.

The example illustrates that the same types of relationships hold for content providers as for hardware manufacturers:

- Moving down the rows of the table, as the portion of users who have upgraded to new technology hardware increases, users are willing to pay a higher price for new technology content, or equivalently, content providers can sell more new technology content at any given price.
- Moving across the columns of the table, if content providers want to sell more new technology content to users, then they must have a larger pool of users buying their content, or they must charge a lower price.

Portion of Content Demanded for New Technology											
	Portion of Content Supplied for New Technology										
Portion of Users with New Technology	1%	10%	20%	30%	40%	50%	60%	70%	80%	90%	99%
1%	0.00	0.00	0.01	0.01	0.02	0.02	0.02	0.03	0.03	0.04	0.04
10%	0.00	0.02	0.05	0.08	0.11	0.14	0.17	0.21	0.24	0.28	0.32
20%	0.00	0.04	0.08	0.14	0.19	0.25	0.31	0.38	0.44	0.51	0.57
30%	0.00	0.05	0.12	0.19	0.27	0.35	0.44	0.53	0.62	0.72	0.80
40%	0.00	0.07	0.15	0.24	0.35	0.45	0.56	0.68	0.79	0.92	1.00
50%	0.00	0.08	0.18	0.30	0.42	0.55	0.68	0.82	0.96	1.00	1.00
60%	0.01	0.09	0.21	0.35	0.49	0.64	0.79	0.96	1.00	1.00	1.00
70%	0.01	0.11	0.24	0.39	0.56	0.73	0.91	1.00	1.00	1.00	1.00
80%	0.01	0.12	0.27	0.44	0.62	0.81	1.00	1.00	1.00	1.00	1.00
90%	0.01	0.13	0.30	0.49	0.69	0.90	1.00	1.00	1.00	1.00	1.00
99%	0.01	0.14	0.33	0.53	0.75	0.98	1.00	1.00	1.00	1.00	1.00

Figure 3-9 User demand for software example: table of portion of content demanded for new system

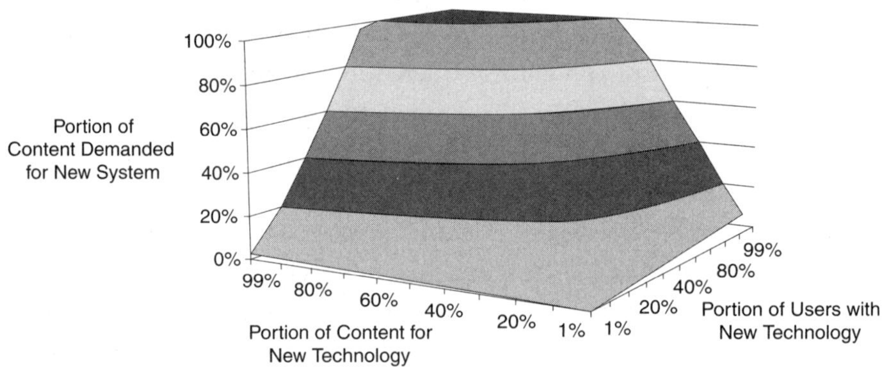

Figure 3-10 User demand for software example: graph of portion of content demanded for new system

3.3 Summary of User Demand for New Systems Hardware and Content

In this chapter I have discussed the various factors that influence users' decisions to adopt hardware and software components for new technology systems. These factors are summarized in Figure 3-11.

Regarding user demand for hardware, I discussed three factors that have a large impact on hardware demand. The first factor is the size of the installed base of users, where the number of existing users can increase hardware demand by new adopters in any of three different ways: (1) through direct network effects, (2) through system validation, and (3) through indirect network effects. I also concluded that the size of the installed base should have a larger impact on increasing new adoption of the system hardware early on, and it should have a diminishing impact on increasing new adoption as the size of the user base grows. The second factor that has a large impact on user demand for hardware is the stock of content available for the technology system. I discussed how some systems are more dependent than others on software for creating value for the system. I also discussed how the impact of content availability on the demand for hardware may increase or decrease as more content becomes available, depending on whether or not the technology system exhibits content saturation. The third factor I discussed that has a large impact on user demand for hardware is the price of new technology hardware and software. I concluded that user demand for hardware will be more sensitive to price than user demand for content, that first-time buyers will be more sensitive to price as adoption progresses from early to late adopters, and that repeat purchasers will be less sensitive to price as risks associated with adoption decrease.

	Nature of Impact on Demand	Magnitude of Impact on Demand
User Demand for New Technology Hardware		
Installed Base	Increases demand through direct network effects and validation.	Impact will initially be large, but it will decrease in effect as more users adopt.
Stock of Content Available	Increases demand by enhancing system value.	Impact may be large or small depending on the system; impact may increase or decrease as more content becomes available, depending on whether later content complements or substitutes for earlier content.
Price of Hardware & Content	Demand is more sensitive to the price of hardware than of content.	Impact on demand increases over time for later first-time buyers; impact on demand decreases over time for repeat buyers as risks abate.
User Demand for New Technology Content		
Installed Base	Increases demand by increasing base of users.	Impact may be large or small depending on the system; impact will tend to decrease over time.
Stock of Content Available	Increases demand by enhancing system value.	Impact may be large or small depending on system; impact may increase or decrease as more content becomes available, depending on whether later content complements or substitutes for earlier content.
Price of Content	Demand is more sensitive to the price of hardware than of content.	Impact on demand increases over time for later first-time buyers; impact on demand decreases over time for repeat buyers as risks abate.

Figure 3-11 Summary of impacts on user demand for hardware and content

In a nutshell, anything that increases the net value of a system will increase the demand for system hardware, where *net value* is the total value provided by the consumption experience, less the price of the system, less any switching costs. When there are fewer alternatives that do what the system does, its net value will be higher, and so, too, will be the demand for the hardware. If a system exhibits direct network effects, then having more users increases the net value of the system to users, so larger installed bases of users will increase demand for the hardware. Likewise, if a system exhibits indirect network effects, so that content availability enhances the value of the system, then having more content available will increase demand for the hardware. If prices for the hardware or software are lower, then the net value provided by the system will be larger, and so, too, will be the demand for the hardware. And finally, if switching costs are lower, then the system's net value will be higher, and along with it the demand for the system hardware.

For software demand, the most important factor is the degree to which the technology system is dependent on software for creating value. Demand for software will generally be high when software greatly enhances the value of the system to users, and it will be small otherwise. More generally, though, I discussed how the same three factors that affect hardware demand also affect software demand: (1) the size of the installed base of users, where content demand will grow faster with the size of the installed base when the technology system is more dependent on software for creating value; (2) the stock of content available for the technology system, where content demand will grow faster, again, when the technology system is more dependent on software for creating value, and also when different units of content tend to complement, rather than substitute for, one another; and (3) the price of new technology software, where users are less sensitive to the price of system software than of hardware, assuming they buy software after they commit to a particular technology system, and where price sensitivities for both hardware and software will increase over time for more durable systems and decrease over time for less durable systems.

It is crucial that system innovators understand these incentives and dynamics of user demand for hardware and software, since they form the basis of the entire Hardware–Software Game. It is, after all, the users' actions and incentives to which hardware manufacturers and software providers must cater and react for them to be able to sell their wares in the marketplace. Without an understanding of these actions and incentives, innovators may as well pack it up and go home, because the odds of winning the Hardware–Software Game are slim enough when the developers know what they are doing. Especially important is the understanding that the value a new technology provides users must be considered within the existing environment, that is, relative to the systems currently being used to accomplish the same or similar things that the new technology system proposes to do. Failure to do so will cause system innovators to overestimate the value their new systems provide and/or the speed at which users will adopt their new systems. Finally, with the increasing importance of networks in today's society, it is also important that new system innovators make their new technologies compatible (in the Everett Rogers sense) with users' existing lifestyles and networks, so as to make transition from older to newer systems as seamless as possible.

CHAPTER FOUR

Technology System Suppliers

In the previous chapter I discussed technology system users, whose actions and incentives lie at the heart of the Hardware–Software Game. The other set of players that I consider are the technology system suppliers of hardware and suppliers of software or content. To understand how the game will play out on its own and/or how the speed of adoption of new systems can potentially be increased, I must understand what factors affect each type of suppliers' decisions to provide new technology components, together with how these factors change in response to actions taken by the other players. These are the issues I take up in this chapter.

Hardware manufacturers sell durable products. So, each user will buy one unit of a hardware manufacturer's product (at least initially during the critical period between new system introduction and through attainment of critical mass). The problem for the hardware manufacturer is to decide when (during which period) to sell each user a unit of hardware. The issue of when is critical because the timing of the sale will determine the price each user is willing to pay for the unit of hardware. This is an intertemporal problem: Hardware manufacturers must figure out the best way to distribute sales of hardware units across periods so as to maximize total profits.

Hardware manufacturers must figure out the best way to distribute sales of hardware units across periods so as to maximize total profits.

In contrast, content providers sell nondurable or consumable products. This means they can sell to each user during each period. During each period there is a different distribution of users across (old and new) technology systems. The problem for content providers is to decide how to divide their content between old and new system users.

The distribution of users across systems will determine the price they are willing to pay for content. This is an intra-temporal problem: Content providers must figure out the best way to distribute content each period across old and new system users so as to maximize total profits.

Content providers must figure out the best way to distribute content each period across old and new system users so as to maximize total profits.

This chapter first discusses the problems and incentives hardware manufacturers face and then discusses the concerns software providers face.

4.1 Provision of Hardware

The term *personal digital assistant* (PDA) was first used in January 1992 to refer to the Apple Newton.[1]

> *Personal digital assistants (PDAs or palmtops) are handheld devices that were originally designed as personal organizers, but became much more versatile over the years. A basic PDA usually includes date book, address book, task list, memo pad, clock, and calculator software. Many PDAs can now access the Internet via Wi-Fi, cellular or Wide-Area Networks (WANs) or Bluetooth technology. One major advantage of using PDAs is their ability to synchronize data with a PC or home computer.*[2]

Handheld calculators have been around since the 1970s; paper-based bound personal organizers became popular in the early 1980s, pioneered by Day Runner;[3] and laptop computers started taking off during the late 1980s. All of these products provided viable alternatives to electronic PDAs when the Apple Newton, and later the Palm Pilot, was introduced into the marketplace during the mid-1990s. A comparison of Apple's experience with the market introduction of its Newton to that of Palm's Pilot

1. Toshiba was arguably first on the scene with an integrated organizer in 1978. Subsequent organizers included the Panasonic Personal Partner, the Atari Portfolio, the HP 95LX, Fujitsu's Poqet PC, and Sharp's PC-3000. www.snarc.net/pda/pda-treatise.htm.
2. www.mobileapps.org.
3. www.fundinguniverse.com/company-histories/Day-Runner-Inc-Company-History.html.

is illustrative of the issues and incentives faced by hardware manufacturers when they bring new technology systems to market.

The Newton was not what the creators had in mind when they first started developing the technology that ended up as the Newton. Rather, the Apple Newton, as it was released in August 1993, was a revamped version of an earlier, more ambitious project that had failed before it ever got to market. The Newton offered the basic features of a PDA, address book, date book, task list, memo pad, and expense log. However, it was also relatively large and heavy for a handheld; it used a revolutionary new operating system whose complexity caused some serious usability problems; its most touted feature, the handwriting recognition system, functioned poorly; and the feature enabling it to synchronize with a Mac or PC did not work well, rendering this capability virtually useless and "making the Newton a data island."[4] Even though subsequent models of the Newton "were vastly improved," the Newton failed to take hold in the marketplace, and Apple eventually halted production of it in 1998.[5]

After seeing the unsuccessful experiences of the Apple Newton and the Palm Zoomer, both of which were released in 1993, Palm discovered the essence of what a successful PDA would be like.

> *. . . during the spring of 1994, Palm conducted in-depth surveys of the hardy souls who had actually purchased Zoomer. What these people said opened the company's eyes. More than 90% of Zoomer owners also owned a PC. More than half of them bought Zoomer because of software (offered as an add-on) that transferred data to and from a PC. These were business users, not retail consumers. And they didn't want to replace their PCs—they wanted to complement them. People weren't asking for a PDA that was smart enough to compete with a computer. They wanted a PDA that was simple enough to compete with paper.*[6]

> *Rob Haitani [product design architect for PalmOne] devised a theory to explain how a handheld could succeed on the market. It was called the Zen of Palm. Any handheld would have to have clear advantages to paper DayPlanners and very few disadvantages. The UI would have to be impeccable. If a user had to navigate multiple dialog boxes, then it wasn't worth the money to get a PDA. A simple benchmark of the efficiency or inefficiency of [a PDA] was to count the number of taps to create an appointment or*

4. http://listing-index.ebay.com/games/Apple_Newton.html.

5. www.wired.com/news/mac/0,2125,54580,00.html.

6. Dillon 1998.

add an entry to the address book. This required that all of the most used features be easily accessible, not buried behind menus or in dialog boxes. This concept of ease of use had eluded many of the early PDA's.[7]

In March 1996, "Palm stepped in and quickly dominated the handheld market with the wildly popular Palm Pilots, which were smaller, cheaper and easier to use"[8] than the Apple Newton. The Palm Pilot offered all the necessities; these features all worked well; and the device was simple and compact:

The PalmPilot 1000 was everything the Apple Newton MessagePad (#28) wanted to be: a "personal data assistant" small enough to fit in your shirt pocket, with enough flash RAM (128KB) to hold a then-impressive 500 names and addresses. The handwriting recognition actually worked (once you mastered the arcane Graffiti software), and best of all, you could sync your data with a PC or Mac desktop application. The brilliance of the Palm concept was its recognition that people wanted a supplement to their computers, not a substitute. Subsequent models grew smaller and more powerful, but were basically refinements to the original PalmPilot's elegant simplicity.[9]

What type of public exposure or word-of-mouth recognition did the two systems receive? Failure of the Apple Newton's handwriting recognition system led to widespread public mockery. For example[10]:

- Garry Trudeau, author of the comic strip *Doonesbury*, ridiculed the Newton in a series of 1993 episodes.
- From an episode of *The Simpsons* titled "Lisa on Ice," which first aired November 13, 1994, Kearney, the school bully, says to his buddy Dolph, "'Hey Dolph, take a memo on your Newton: beat up Martin.' [Dolph writes "Beat up Martin" which the Newton translates as "Eat up Martha"] 'Bah!' [throws Newton]."
- Ridicule of the handwriting recognition led to the joke "How many Newton users does it take to change a lightbulb? Foux. There to eat lemons, axe gravy soup."

The Palm Pilot, on the other hand, benefited from tremendously positive publicity generated by two awards it received, *PC Computing*'s MVP "Usability Achievement of

7. Hormby 2007.

8. www.wired.com/news/mac/0,2125,54580,00.html.

9. www.pcworld.com/reviews/article/0,aid,123950,pg,2,00.asp.

10. www.spiritus-temporis.com/apple-newton/appearances-in-popular-culture.html.

the Year" Award during the fall of 1996 and *Newsweek*'s "High Tech Gizmo of the Year" award in December of 1997.[11]

Even the prices of the two systems greatly favored Palm over Apple. Earlier Newton models were slightly less expensive than the later models, $500 to $800 versus about $1,000. At $299 to $399, or about one-third to one-half the price of Apple Newtons, early models of Palm Pilots were relatively inexpensive. The fact that the Palm Pilot was simpler and easier to use than the Newton, not to mention the fact that its features actually worked, also meant that the time costs associated with the use of Palm Pilots were undoubtedly less than those experienced by Newton users.

"During product development, Donna, Jeff and Ed [the founders of Palm] were traveling the country promoting the Touchdown as the platform of choice for hardware and software developers"[12] to create third-party applications. As early as 1998, a mere two years after it was first introduced into the marketplace, the Palm Pilot had spawned an entire ecosystem.

> *Today the PalmPilot is more than a hot product. It's the center of an industry. More than 5,000 programmers are working on new software applications. Some 200 developers are designing hardware add-ons. And the Palm PDA is more than a business phenomenon. It's a cultural phenomenon . . .*[13]

The Newton OS, NewtonScript, was

> *an easier to use syntax based upon the Self language. It still pushed the state of the art in object oriented coding, automated memory management, and what is today referred to by Microsoft as managed code ... The Newton's object oriented system also allowed plugins to add dramatic new functionality to existing applications.*[14]

So Apple clearly embraced the contributions of third-party suppliers.

However, although subsequent availability of applications and accessories provided by Apple, Palm, and other third-party suppliers increased the value of the Apple Newton and Palm Pilot systems for later adopters, indubitably it was the out-of-the-box functionality, or lack thereof, that contributed to the eventual failure of the Apple Newton (and Palm Zoomer) and the success of the Palm Pilot. Figure 4-1 provides a

11. www.palm.com/us/company/corporate/timeline.html.

12. Hormby 2007.

13. Dillon 1998.

14. Eran 2006.

Model [A]	Release Date [B]	Discont. Date [C]	Price [D]	Wt (oz.) [E]	Size (in.) [F]	Memory (MB) [G]
Apple						
MessagePad (OMB)	Aug-93	Mar-94	$700	14.4	7.25 x 4.5 x 0.75	4
MessagePad 100	Mar-94	Apr-95	$500	14.4	7.25 x 4.5 x 0.75	4
MessagePad 110	Mar-94	Apr-95	$600	20.5	8.0 x 4.0 x 1.25	4
MessagePad 120	Jan-95	Jun-96	$600	16.0	8.0 x 4.0 x 1.25	4
MessagePad 130	Mar-96	Apr-97	$800	16.0	8.0 x 4.0 x 1.25	8
MessagePad 2000	Mar-97	Feb-98	$950	22.4	8.3 x 4.7 x 1.1	4–8
MessagePad 2100	Nov-97	Feb-98	$1,000	22.4	8.3 x 4.7 x 1.1	8
Palm						
PalmPilot 1000	Mar-96		$299	5.7	4.7 x 3.2 x 0.7	0.256
PalmPilot 5000	Mar-96		$369	5.6	4.7 x 3.1 x 0.7	0.512
PalmPilot Personal	Mar-97		$299	5.6	4.7 x 3.1 x 0.7	0.512
PalmPilot Professional	Mar-97		$399	6.0	4.7 x 3.2 x 0.7	1
Palm III	Mar-98		$399	6.0	4.7 x 3.2 x 0.7	2

	Apple	Palm
Handwriting Recognition	Initially Poor	More Robust
Sync with Mac/PC	Complex	Easy

Sources:

www.businessweek.com/1996/37/b3492103.htm
www.businessweek.com/1997/27/b353441.htm
www.businessweek.com/1998/12/b3570044.htm
www.byte.com/art/9607/sec13/art6.htm
www.palmzone.net/modules.php?name=Pda_Guide&rop=list&show=All_Models
www.palmblvd.com/hardware.html
www.everymac.com/systems/apple/messagepad/index.html

Figure 4-1 Early PDA specifications: Apple Newton versus Palm Pilot

summary of the salient characteristics of the early Apple Newton and Palm Pilot models, and Figure 4-2 provides a side-by-side photograph of the two types of PDAs.

What the experiences of Apple with its Newton and of Palm with its Pilot tell me is that a new technology system's hardware will live or die before the technology ever hits the market. The technology must be designed to meet the users' needs as simply, compactly, and cheaply as possible. Serious technology malfunctions or overly ambitious designs will almost invariably doom a technology to failure.

Now that I have examined the specific case of PDAs, suppose more generally that you are a system innovator, and you have just introduced the hardware component for a new technology system into the marketplace. You know this may take some time, but

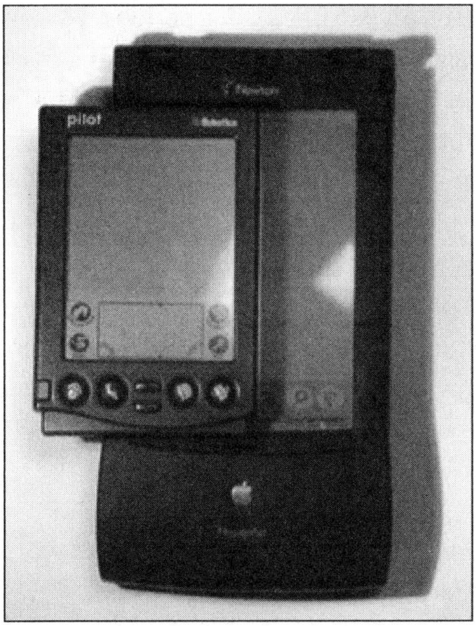

Figure 4-2 Apple Newton versus Palm Pilot. (Photo reprinted with permission from Casimir Decas.)

the goal is to eventually get the technology adopted and earn as much money as possible so you can support your serial entrepreneurship addiction. What you have to do is to figure out how to distribute sales of hardware units to users over time; that is, you must choose prices for your hardware units each period so as to encourage the optimal pattern of hardware adoption by users. The optimal pattern of adoption is that which will earn you the greatest amount of profits over time.[15]

Your choice is a dynamic one, because the technology system's direct and/or indirect network effects mean that having a larger installed base of users during earlier periods will increase the value of the technology system to later adopters, and therefore the price they will be willing to pay. This means that you must consider not just current profitability when trying to decide current actions. Rather, you must also consider the impact on *future* profitability that your *current* actions will have. In fact, you may sometimes

15. Theoretically, manufacturers can either choose the price to set for their hardware and then the market will determine how many units will be sold at that price, or they can choose the number of hardware units to supply and the market will determine the price that will lead users to buy that number of units. I believe the former is a more realistic picture of what actually happens in the real world.

be better off sacrificing profits during earlier periods, by selling "too many" units at "too low" a price, to generate much larger profits during later periods.

Interconnectedness issues also have a large impact on the hardware manufacturer's choice. Specifically, profitability will be higher when there is a lot of content available for the new technology system. Unfortunately, though, the hardware manufacturer has no control over content availability. Rather, it is the content providers that decide how much content to make available. Your job, then, is to induce content providers to provide as much content as possible.

So now as a manufacturer of new system hardware, you face the problem, expressed in Figure 4-3, of choosing hardware prices each period. What determines the prices you choose and whether or not users will buy your system hardware at those prices?

First note that regarding costs per unit, **Unit Cost**$_t$, the higher your costs are, the lower your profits will be, for two reasons. First, when your costs are higher, your profitability—that is, price less cost—will be lower whatever price you charge. And second, higher costs mean you have to charge a higher price to cover your costs, and higher prices mean lower sales volume, and thus lower profits. Either way, higher costs lead to lower profits, which means you generally want to engineer your hardware to have the lowest costs possible. Palm did this; Apple did not.

> System hardware should be engineered to have the lowest unit costs possible.

The next variable to consider is number of units sold each period, **Units**$_t$, which you know from the previous chapter will depend on (1) the price you set for the hardware, where higher prices mean fewer sales, except for those few users for whom the value of the technology hardware is greater than the high price you set; (2) the size of

Goal: Choose prices each period t, Price$_t$, to maximize total profits:

$$\text{Total Profits} = \sum_{t=0}^{\infty} \text{Profits}_t$$
$$= \sum_{t=0}^{\infty} (\text{Price}_t - \text{Unit Cost}_t) \cdot \text{Units}_t$$

Figure 4-3

the installed base, where larger installed bases enable greater direct network effects, as well as validation for the technology system, both of which will enable you to sell more units; and (3) the availability of content, where greater availability makes the system more valuable to users, in which case you will sell more units. Palm's lower costs enabled it to set lower initial prices for its PDAs than could Apple. Also, Palm's features were simple, yet elegant and functional, leading to high value and positive word-of-mouth accolades (strong validation), which enabled Palm to quickly build up a large base of users. In contrast, early Apple users experienced undue complexity and problems with the Newton, leading to low value and public ridicule. Each of these issues led to greater unit sales of Palm Pilots and fewer unit sales of Apple Newtons.

The last and most complex variable to consider is the price to charge each period, **Price$_t$**. For systems that exhibit scale economies and network effects, generally speaking, you should set lower prices during earlier periods than you might otherwise be inclined to charge. Lower prices mean higher sales volume. As you sell more hardware units early in the technology's lifecycle, (1) your per-unit costs of manufacturing new technology hardware drop over time, due to learning-by-doing effects, technology advances, and economies of scale; and (2) you will generate a large installed base of users earlier, which as you know from the previous chapter will increase later sales through direct network effects, validation, and indirect network effects.

Furthermore, from the previous chapter you know that users will find your technology hardware to be more valuable, and thus enable you to charge a higher price, when your new technology system satisfies one or more of the following three conditions.

First, your technology hardware will be more valuable when it can do what it has been designed to do better than existing alternatives can. The more similar your new system is relative to current alternatives, the more the prices of those alternatives will limit the price you will be able to charge for your new system. Your new system must provide at least some minimum threshold of value over current systems to get users to incur the costs of switching. And the more revolutionary your system is compared to what is already out there, the greater the price you will be able to charge. Palm's simplicity, elegance, and functionality provided sufficient value to users; Apple's unwieldiness, complexity, and dysfunction did not.

Second, your technology hardware will be more valuable when there is more content available for the new system. This means that you want third-party providers of content and accessories to provide as much content as early as possible for your new technology. The more content that is available, the greater will be the value of the new system to users, and so the more they will be willing to pay for your new system hardware. It appears that Palm spent more effort trying to get third-party developers on board than Apple did, probably because Apple had a firm reputation that attracted developers with less effort than the less well-known Palm. However, as indicated previously, in the case of PDAs, providing a system with good out-of-the-box functionality was probably the most important factor contributing to the success of Palm, which then led to a quickly created ecosystem, as opposed to having many third-party applications available upon release of the new system.

And third, your technology hardware will be more valuable when there are more adopters of the new system as early as possible. Having a large installed base of users for your technology system will be more important for increasing the appeal of your hardware when direct network effects create a lot of value for users of your new technology system, and/or when early adopters are a more important source of validation for your technology.

> System hardware will be more valuable to users and thus garner higher prices when
>
> - It is much better at doing what it's supposed to do than current alternatives
> - There is a large amount of content available
> - There is a large installed base of users

A further consideration is that when large installed bases of users create a lot of value for future adopters, there is the potential for tension regarding the optimal volume of current versus future sales of hardware: Adopters become more sensitive to price as the lifecycle moves from earlier to later adopters. This dynamic will tend to encourage you to sell to small numbers of users early on at high prices, and then decrease the price over time to attract more price-sensitive adopters during later periods. With this type of price-skimming strategy (that is, skimming off the high-paying users before lowering the price to users with lower willingness to pay) you effectively sacrifice profits tomorrow in exchange for higher profits today (see Figure 4-4). When early adopters are less sensitive to price and network effects are not as strong, hardware manufacturers will tend to do better by initially selling to only a few early adopters,

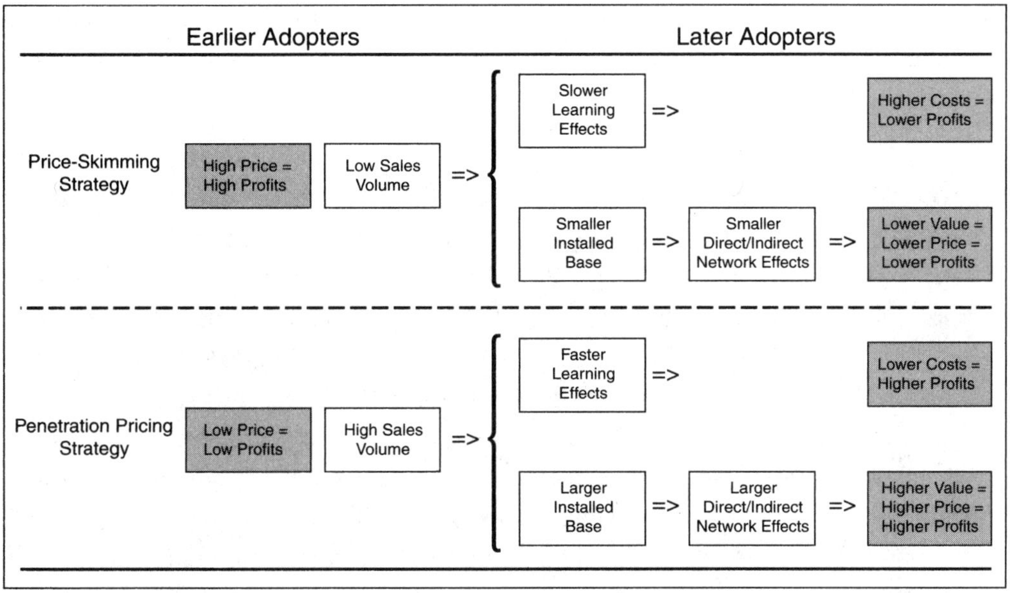

Figure 4-4 Price-skimming versus penetration pricing strategies

earning large margins, and then subsequently dropping the price to attract users who are more sensitive to price.

Alternatively, lower prices today mean lower margins on today's sales, but more sales today, and so larger installed bases and thus more value (and higher prices) to users tomorrow. With this type of penetration pricing strategy, you sacrifice profits today in exchange for higher profits tomorrow (see Figure 4-4). The smallest sacrifice for the largest gain occurs when early adopters are relatively more sensitive to price and when direct network effects are large. In these cases, hardware manufacturers will tend to do better by selling more units during the early periods at a lower price, which will then enable them to charge a much higher price during the later periods.

Taken together, these dynamics suggest that the more the value of your new technology system is dependent upon direct network effects—that is, a large installed base of users—the greater will be your tendency to choose a penetration pricing strategy, rather than a price-skimming strategy, so as to build up a large installed base of users as quickly as possible to create more value for future adopters.

Finally, there is a tendency for hardware manufacturers to want to achieve critical mass as soon as possible so as to increase their probability of long-term success in the marketplace. This desire to achieve critical mass will tend to lead hardware manufacturers to set lower prices during the early stages of adoption so as to generate greater levels of system adoption.

In conclusion, the specific case of PDAs and the more general discussion of the introduction of a generic technology system expose two distinct considerations for hardware manufacturers seeking to introduce new technology systems into the marketplace: (1) how to engineer a new technology system so it is most likely to be adopted, and (2) how to introduce the engineered technology into the marketplace so it is most likely to be adopted. In the case of PDAs, we saw that successfully engineered technologies provide value to users above and beyond that offered by existing alternatives, they function as they are supposed to, they work with the existing infrastructure and/or within the existing environment, they are simple to use, and they are provided at a low price. Innovators of new technology systems that are successfully introduced into the marketplace recognize the importance of technology value and manage to exploit direct and indirect network externalities appropriately.

4.2 Provision of Content

Content providers sell nondurable or consumable products, which means they can sell to each user during each period. The problem for content providers is to decide how to divide their content between old and new system users. The distribution of users across systems will determine the price they are willing to pay for content. This is an intra-temporal problem: Content providers must figure out the best way to distribute content each period across old and new system users so as to maximize total profits.

At the time the Apple Newton and later the Palm Pilot were being developed, third-party software providers were supplying content for various types of computer systems (personal computers, workstations, gaming systems, and so forth) for which software and accessories enhance the user experience. Generally speaking, the markets these third-party developers were serving had installed bases of users of older technology systems; that is, they were supplying markets with an established demand for their products and services.

When news of the development of the Newton and Palm Pilot spread, these third-party suppliers had to decide if and when to switch their resources from the production of content for the older technology systems to the production of content for the newer technology systems: Continue to supply established, proven-to-be-profitable markets, or switch over to supplying a new, uncertain, may-take-off-or-may-die-out market? Of course, there is already plenty of competition in the more established markets, so profits, while relatively certain, are also generally limited. The new markets, on the other hand, while uncertain, hold the promise of potentially greater profits if they do eventually take off. Content providers' profitability associated with old and new technology system users thus depends on the quantities of content sold to the populations of old and new technology system users, as well as on differences in margins achieved for each group.

Four issues are important in determining the speed at which third-party providers will transition from older, more established markets into newer markets. The first issue involves content saturation, defined broadly. If users of an existing technology system are reluctant to purchase new content because they already have "enough" to satisfy their needs, then the market for that system has become saturated. In this case, content providers will be inclined to transition more quickly to new technology markets, since demand for old system content has dried up. For example, the sales of new DVDs have been decreasing, because the market has become saturated.[16] Correspondingly, system providers have been eager to introduce the next-generation system, recently determined as Blu-ray, so users will once again start buying content.

Conversely, suppose that content providers introduce content for the new system, content that old system users cannot access with their older technology system hardware but would have bought had they been able to use it. In this case, the old system market for content has not yet been saturated, and sales to new system users come at the expense of sales to old system users; that is, content providers are facing product cannibalization. When existing markets have not yet been saturated, content providers would do better by transitioning more slowly and selling more content to users of the older system. Later, when the older market has become saturated, content providers may then be able to sell the same content again, now upgraded for the new technology, to those same users who have later upgraded to the new system.

16. See, for example, *Dealerscope* 2007.

> Content providers will be inclined to transition more quickly to new technology markets when existing markets have been saturated.

The second issue influencing the speed of content provider transitioning involves the level of uncertainty surrounding the new technology system. The greater the uncertainty about if and when the new market will become established is, the slower the transition by content providers out of older markets and into newer ones will be. This problem plagued the HD market for a long time. The format wars pitting HD-DVD against Blu-ray left users and content providers standing on the sidelines, waiting for one of the two formats to prevail, before they would pick a side. The war was finally resolved in favor of Blu-ray, at which point sales began to take off.

> Content providers will be inclined to transition more slowly when there is a lot of uncertainty about if and when the new market will become established.

The third issue affecting the speed of content provider transitioning has to do with user groupings. It is often more profitable for content providers to have all users in one place—that is, either everyone has the old technology system or everyone has the new technology system—rather than having users divided between different systems. More specifically, even though the new technology system content may be potentially more profitable for content providers once all users have upgraded to the new system, total profits may very well be higher when most users have the old technology, as opposed to having some with the old and some with the new. When there is some type of fixed cost associated with providing each type of content to the different groups of users—that is, there are economies of scale in the production of content—profits per user may very well be lower when users are divided, rather than unified (see Figure 4-5). In this case, content providers may be slow to transition from the old system to the new one, if profits per user are larger in the older market, until a certain threshold of users has already transitioned.

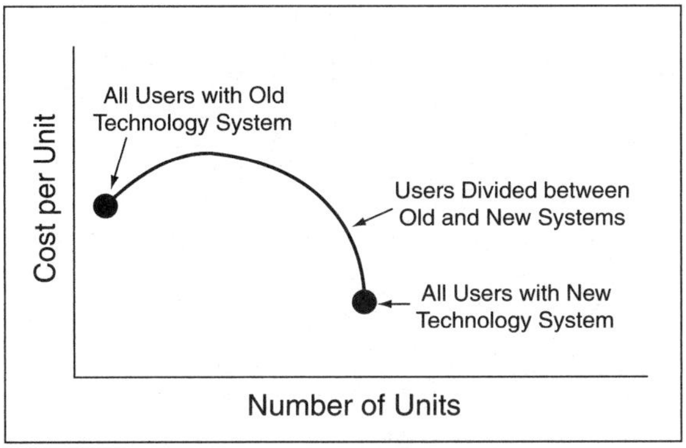

Figure 4-5 Cost per user in unified versus divided markets

> Content providers will be inclined to transition more slowly when there are large fixed costs associated with providing content for each group of old and new technology users.

This problem is currently plaguing media advertisers. A few decades ago, consumers spent a large portion of their spare time watching television. The problem for advertisers then was to decide during what time and on which channel to broadcast their advertisements. Currently, however, the attention of consumers is much more dispersed. Much of the time they used to spend watching television is now allocated to surfing the Internet and/or playing video games. Consequently, any given broadcast advertisement, be it tied to television, the Internet, or video games, is not reaching as many consumers as earlier advertising that was broadcast when more consumer attention was focused on television. So even though it may currently cost advertisers only a portion of the cost to broadcast an ad on the Internet as it used to cost to broadcast the same ad on television, dispersion of user attention means the ad has less reach, and thus higher costs per viewer.

The fourth and last issue affecting the speed of content provider transitioning involves the cost of transitioning from providing content for the old system to providing content for the new system. Up-front or fixed costs of transitioning from the older to the newer market include, for example, costs associated with buying new equipment, learning new methods, and lost sales during the transitioning downtime. The larger the costs of transitioning are, the slower the transition by content providers out of older markets and into newer ones will be. The problem of large costs of investing in new equipment is hampering the transition of movie theaters into providing digital projections:

> *Studio executives and theater owners agree on the merits of showing crisp, clean digital images to moviegoers, the majority of whom are already tuned in to a variety of new media. But they're still struggling with how to handle the cost of converting to new equipment.*[17]

> Content providers will be inclined to transition more slowly when there are large costs associated with transitioning from providing content for the old system to providing content for the new system.

To summarize, four issues important in affecting the speed of content provider transitioning from older to newer technology systems are (1) content saturation, where lower levels of saturation in older markets, or the possibility of product cannibalization, decrease the speed of provider transitioning; (2) uncertainty, where greater levels of uncertainty regarding if and when a new system will take off decrease the speed of transitioning; (3) profit per user, where higher profits associated with serving older technology markets decrease the speed of transitioning; and (4) costs of transitioning, where higher up-front costs decrease the speed of transitioning.

As we see, certain market conditions will tend to dampen the speed of transitioning for content providers from older to newer technology systems. However, as mentioned in section 4.1, this creates a point of contention with hardware manufacturers. Hardware manufacturers want the service providers to enter the market and supply as much content as possible for the new technology system precisely at that time when it is least profitable for content providers to do so: before the technology has been established and many users have adopted the new technology. At this point, having more content will help the hardware providers establish the value of their product, yet during this time there is no assured value of which the content providers can avail themselves.

17. Cole 2007.

CHAPTER FIVE

The Hardware–Software Game

5.1 Introduction to Game Theory

I once heard a venture capitalist say that one of the most frequent mistakes entrepreneurs make when they pitch their ideas to him in the hopes of receiving funding is that they do not account for reactions by competitors to the start-ups' entries into the market. Anyone who has benefited from the price wars instigated when Southwest or JetBlue has entered one of American's or United's airline routes knows that incumbent response to new entry is a very real phenomenon indeed. The failure of new entrants to anticipate and account for reactions by existing firms that want to protect their positions can easily lead to the start-ups' quick defeat and exit from the market.

The Hardware–Software Game takes place in just such a setting. When hardware manufacturers introduce next-generation technology systems into the marketplace, content providers are there who have been serving users of older technology systems. The manufacturers of next-generation hardware would like content providers to immediately start supplying content for their new systems. Yet manufacturers must realize that content providers have a very real interest in protecting their positions in the markets for established systems.[1] If suppliers of new system hardware fail to understand the interests and incentives of content providers vis-à-vis users of established technologies, and if they fail to specifically address these considerations, they will jeopardize their chances of attaining critical mass in the markets for their new technology systems.

Game theory provides us with a means of analyzing the structure of the Hardware–Software Game. This type of analysis helps us better understand the incentives of hardware manufacturers, content providers, and users within the context of the game. From

1. Of course, other hardware manufacturers, providing old technology hardware, may also be in the market, and they, too, have an interest in protecting their positions. However, we may simplify the discussion by excluding these suppliers without hurting the results of the analysis.

these insights we can more easily figure out how to influence the players' incentives so as to increase the odds of successful adoption of new technology systems.

Accounting for market reactions to any given player's actions is crucial. The increasing degree of interconnectedness in today's markets means that actions by any one player can affect other market participants more than ever before. It follows that for market entrants and participants to make the right decisions, they must understand two things in particular: (1) how their actions affect other players, and (2) what the other players' options and expected reactions will be. Only by optimizing those sets of interconnected actions and reactions as a whole will market participants maximize their chances of succeeding in the marketplace.

Game theory provides a means of capturing interactive decision problems. In a game, several players attempt to optimize their own payoffs, while explicitly taking into account how the actions of the players affect the payoffs to each player. More precisely, the definition of a *game* is (1) a description of each of the players involved, (2) the options or actions available to each of them, and (3) the payoffs to each player associated with each set of potential actions taken by all the other players. A *non-cooperative Nash equilibrium* is a set of *stable* or *self-reinforcing* actions. A set of actions is stable if, given the actions of all the other players, no one player finds it in his or her own best interest to choose any action other than the equilibrium action. In other words, an equilibrium is a set of actions that are in every player's best mutual self-interest, given that each player takes into account the presence of the other players and then acts in his or her own best self-interest, independently from all the other players.[2]

The definition of a *game* is

1. A description of each of the players involved
2. The options or actions available to each of them
3. The payoffs to each player associated with each set of potential actions taken by all the other players

2. An alternative to the concept of Nash equilibrium is that of Pareto optimality. A Pareto-optimal outcome is a set of actions for each player such that no other set of actions will leave one player better off without making another player worse off. In other words, the concept of Pareto optimality seeks to find those actions that will maximize the size of the total pie, whereas the concept of Nash equilibrium seeks to find those actions that are self-reinforcing. Some Nash equilibria may not be Pareto-optimal, and some Pareto-optimal equilibria may not constitute Nash equilibria.

> An *equilibrium* is a set of actions that are in every player's best mutual self-interest, given that each player takes into account the presence of the other players and then acts in his or her own best self-interest, independently from all the other players.

Take, for example, the classic Prisoners' Dilemma Game (PDG). Two crooks, Larry and Linda, who committed a crime together, are arrested, taken into custody, and questioned separately from one another, so Larry does not know what Linda is saying, and Linda does not know what Larry is saying. During the questioning, Larry and Linda have the option of either keeping quiet or ratting out their partner. If, acting independently from one another, both choose to keep quiet, then both will receive moderately light sentences of four years of jail time. If, on the other hand, both rat on their partner, then both get a moderately harsh punishment of eight years in prison. Alternatively, if one of the two keeps quiet, while the other rats out his or her partner, then the fink gets a light sentence, two years in jail, and the person who has been sold out faces a long sentence of 12 years in jail.

The basis for the game is the fact that whatever Larry does—keeps quiet or rats out Linda—will affect what happens to Linda—that is, the length of time she will spend in jail—and vice versa. So we have the definition of the game as displayed in Figure 5-1.

Players
- Linda
- Larry

Potential Actions
- Larry: Keep Quiet or Rat Out Linda
- Linda: Keep Quiet or Rat Out Larry

Payoffs

		Larry	
		Keep Quiet	Rat Out Linda
Linda	Keep Quiet	4 Years 4 Years	12 Years 2 Years
	Rat Out Larry	2 Years 12 Years	8 Years 8 Years

Figure 5-1 Prisoners' Dilemma Game

Now let's see what happens. From Linda's perspective, Larry can either keep quiet or rat Linda out. If Larry keeps quiet (the first column in the payoff matrix in Figure 5-1), then Linda faces either four years in prison if she keeps quiet, or two years in prison if she rats out Larry. So if Larry keeps quiet, it is in Linda's best interest to rat out Larry. If, on the other hand, Larry rats out Linda (the second column in Figure 5-1), then Linda faces either 12 years in prison if Larry keeps quiet or eight years in prison if he rats out Linda. So, if Larry rats out Linda, then it is still in Linda's best interest to rat out Larry. From Linda's perspective, then, whatever Larry does, Linda is better off choosing to rat out Larry. Linda's equilibrium strategy is therefore to rat out Larry.

In the same manner, we see that from Larry's perspective, whatever Linda does, Larry is better off choosing to rat out Linda. So Larry's equilibrium strategy is also to rat out his partner:

Nash Equilibrium for PDG

Larry: Rat Out Linda

Linda: Rat Out Larry

The dilemma here is that both Larry and Linda would be better off if they could cooperate and both decide to keep quiet. However, both Larry and Linda keeping quiet is not an equilibrium (not stable), because if Linda decides to keep quiet, then it is in Larry's best self-interest to rat her out, and vice versa. So even though they would prefer that both keep quiet, separately each does best by ratting, so both end up ratting.

A more realistic and more complex game is the Battle of the Sexes. In this game Larry and Linda are married, Linda is at home cooking dinner, and Larry has to stop on the way home from work to pick up a bottle of wine for dinner. Linda can make either steak or salmon for dinner. She knows Larry prefers steak, but she prefers salmon. Larry knows that Linda is making either steak or salmon, and he knows that she prefers salmon, but he likes steak. Both Linda and Larry would prefer to have red wine with steak and white wine with salmon. Of course, this game was much more credible ten years ago before Larry could just whip out his cell phone and find out for sure if Linda is making steak or salmon, but we assume that they cannot contact each other to find out what's what.

Now let's see what happens. From Linda's perspective, Larry can buy either red wine or white wine. If Larry buys red wine (the first column in the payoff matrix in Figure 5-2), then Linda will rate the meal an 8 if she makes steak or a 6 if she makes

Players

Larry

Linda

Potential Actions

Larry: Buy Red Wine or Buy White Wine

Linda: Prepare Steak or Prepare Salmon

Payoffs

		Larry	
		Red Wine	White Wine
Linda	Steak	8 10	4 6
	Salmon	6 4	10 8

Figure 5-2 Battle of the Sexes Game

salmon. So if Larry buys red wine, it is in Linda's best interest to cook steak. If, on the other hand, Larry buys white wine (the second column in Figure 5-2), then Linda will rate the meal a 4 if she prepares steak or a perfect 10 if she prepares salmon. If Larry buys white wine, then, it is in Linda's best interest to prepare salmon. In this case Linda does not have an option that would be a clear winner regardless of what Larry does. She has to decide what to do based on what she thinks Larry will do, that is, what she thinks the probability is that Larry will buy red wine versus white wine.

In the same manner, we see that from Larry's perspective, he also does not have a clear winning choice. He also must choose an action based on probabilities and face the payoff that results when they see what the final outcome is.

Just as in the PDG, in the Battle of the Sexes Game the payoff to Linda depends on what Larry does, and the payoff to Larry depends on what Linda does. Also, the players in both games face a similar dilemma: Both Larry and Linda would be better off if they could coordinate their actions, and both will possibly end up worse off for lack of being able to coordinate. So the outcome of a game is characterized by the fact that (1) the payoffs to each player in the game depend on the actions each of the other players takes, and (2) players would be better off if they could coordinate their actions.

The outcome of a game is characterized by the fact that

- The payoffs to each player in the game depend on the actions each of the other players takes
- Players would be better off if they could coordinate their actions

To reiterate, the two game examples illustrate how each player's actions affect the other player's payoffs, which is exactly the situation that game theory attempts to capture. In today's technology markets, where every market player's actions affect every other market player's payoffs (profits), the use of game theory proves to be an unparalleled method for capturing the underlying dynamics of the market; that is, we can use game theory to better understand the incentives and options faced by each market player: how each market player's actions affect each of the other players, what is in each player's best interest to do, given the actions of the other players, and therefore, what the mutually compatible (equilibrium) outcome will be.

5.2 Definition of the Hardware–Software Game

Now that I have described how a game works, let's look at the Hardware–Software Game from the perspective of game theory. The Hardware–Software Game is a multiperiod game. It starts during the period (month or year) in which a new technology system is introduced into the marketplace, and it ends at the earlier of when the market for the new technology system either (1) becomes saturated or (2) is usurped by yet another technology system. The players of the Hardware–Software Game are users; hardware manufacturers (HW); and software, content, or accessory providers (SW). Each period, hardware manufacturers must decide on the price to set for the new technology hardware, software providers must decide on the portion of their resources to allocate to the production of old and new technology content, and users must decide whether or not to adopt the new technology system by purchasing the hardware. So far for the Hardware–Software Game:

Players

HW

SW

Users

Potential Actions Each Period t

HW: Choose price, $\mathbf{P}_t$, for new technology hardware for each period **t**.

SW: Choose portion of content, $\mathbf{d}_t$, to provide for the new technology for each period **t**.

Users: Decide whether or not to upgrade to the new technology system during period **t**.

Hardware manufacturers set the price of hardware, and users decide whether or not to adopt the new system at that price. When the price is higher, and everything else is the same (namely, content availability and the size of the installed base), fewer users will choose to adopt. In essence, then, hardware manufacturers choose the portion of new users who adopt by setting the price accordingly; that is, hardware manufacturers choose higher prices and fewer adopters versus lower prices and more adopters.

When content availability and the size of the installed base are both held constant, there is thus an inverse relationship between prices and portion of users; that is, if we see that prices are higher, we can expect that fewer users will adopt. And the converse is also true, that if we see that fewer users have adopted, we can infer that prices are higher. In the discussions that follow, I attribute lower adoption rates to higher prices and vice versa.

The last piece of the definition of the Hardware–Software Game is the payoff matrix that shows how much each player gets under alternative actions taken by that player and by the other players. Of course, as well as being affected by the actions each player takes during each period, the payoffs to the respective players will also vary across technology systems, depending upon the market characteristics of each system. In Chapters 3 and 4, I discussed the primary market and technology characteristics that will affect the payoff levels for each of the players (value for users and profits for HW and SW) and thus drive them to choose certain actions over others. The factors I discussed were (1) technology value (that is, functionality) of old versus new technology systems, (2) price sensitivities of users, (3) the sizes of installed bases for old versus new technology systems and the importance of direct network effects in creating value for technology systems, (4) the amount of content available for old versus new technology systems and the importance of indirect network effects in creating value for technology systems, and (5) whether later content complements or substitutes for earlier content, that is, the tendency for content saturation.

I created a mathematical model (presented in Appendix A) that incorporates all these variables and simulates the structure of the problems facing each set of players in the Hardware–Software Game. By using this model, I was able to vary the sizes of the parameters to see how different scenarios lead to different outcomes of the game. Before moving on to the outcome, I will first describe in more detail the assumptions I made about each of the four drivers, which are summarized in Figure 5-3.

Driver	Description	Assumptions
Technology Value = Market Size	Technology systems with greater value (functionality) have greater market sizes (more users will potentially adopt)	Technology system market sizes increase • over time; • with successive generations.
Users' Sensitivity to Price	The degree to which fewer users will adopt the system when its price goes up	Risk and uncertainty cause users to be • more sensitive to the price of newer technology systems; • less sensitive to the price of any given technology system over time.
Magnitude of Direct Network Effects (DNE)	The degree to which interaction with other users increases a technology system's value	The magnitude of DNE • are similar from one generation to the next • decrease over time
Magnitude of Indirect Network Effects (INE)	The degree to which the availability of software, content, and accessories increases a technology system's value	The magnitude of INE • are similar from one generation to the next • decrease over time
Degree of Content Complementarity	The degree to which earlier content complements (increases demand) or substitutes for (decreases demand) later or more content	The degree of content complementarity • is similar from one generation to the next • decreases over time

Figure 5-3 Summary of market drivers

5.3 Assumptions about Market Dynamics

5.3.1 Technology Value and Market Size

As I previously discussed in Chapter 3, early generations of cell phones were large and bulky, had poor coverage, and both phones and service were expensive. In other words, the value of cell phone technology was relatively low, leading to small potential market size, maybe a couple of percent of the population. However, since they were first introduced into the marketplace during the 1980s, cell phones have evolved tremendously, to the point that they are now small; they easily fit on your belt or in your pocket; coverage is generally national, if not international; and service is affordable. It follows that both the technology values and accompanying market sizes of today's cell phones are much greater than they used to be, and, in fact, over 80 percent of the now 300 million Americans currently use cell phones.[3]

This illustrates how market size (or, synonymously, "market potential") and technology value go hand in hand. The more valuable a new technology is—that is, the more superior its functionality is to currently available alternatives—the greater its market size will be. Think of market size and technology value as the portion of a given population that will potentially adopt a new technology. In this case, whereas the market potential of early cell phone technologies was perhaps a few million, the market potential of current technologies is close to the full size of the population.

> Market size and technology value go hand in hand: The more valuable a new technology is—that is, the more superior its functionality is to currently available alternatives—the greater its market size will be.

By definition, technology systems with larger market potentials will tend to attract more users and more third-party suppliers of services and accessories. Over time as the market evolves, new and different applications for technology systems are discovered, which further increases the sizes of the markets. At the same time, learning by doing and other general improvements in technology atmospheres lower the costs of production. Simultaneously, competition forces suppliers to pass the lower costs of production on to users in the form of lower product prices.

3. www.ctia.org/media/industry_info/index.cfm/AID/10323.

The evolution of the Apple iPod Classic provides a perfect example of how prices fall over time with successive generations of technology systems. The first-generation Apple iPod Classic was released on October 23, 2001, with a capacity of 5GB and a launch price of $399. Between October 2001 and September 2007, Apple released 18 different models. Although prices have increased nominally over time, capacity has increased faster, causing the price per unit of capacity to drop exponentially from $80 per gigabyte in October 2001 to just over $2 per gigabyte in September 2007, as seen in Figure 5-4.[4]

Lower production costs enable lower prices. In addition, though, as production costs decrease, technology systems become economical to use for other applications and/or in new markets. For example, when the technology for digital photography was first introduced, only professional photographers could afford the specialty cameras. Now, the technology for digital cameras has evolved to the point that digital cameras are compact enough and inexpensive enough to be included in run-of-the-mill cell phones. So the potential market for digital photography has now grown to include not only that for stand-alone cameras, but that for cell phones as well. Market potentials for next-generation technology systems thus increase as new applications are found.

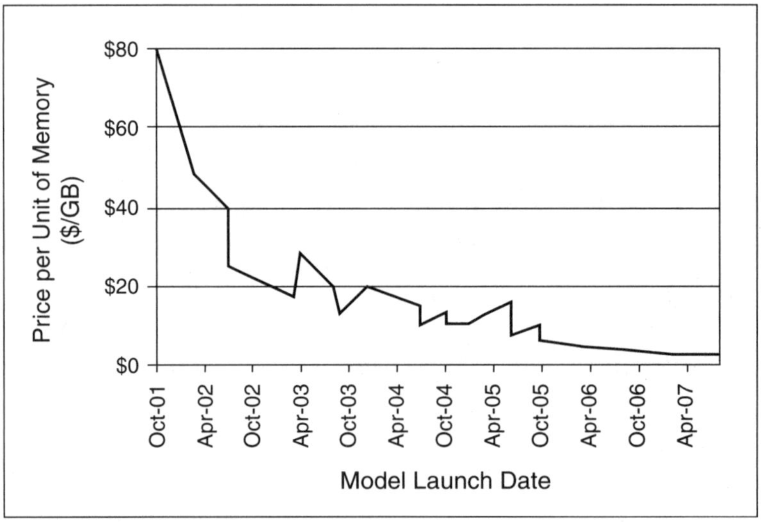

Figure 5-4 Price of memory in Apple iPod Classics

4. http://en.wikipedia.org/wiki/Ipod_classic.

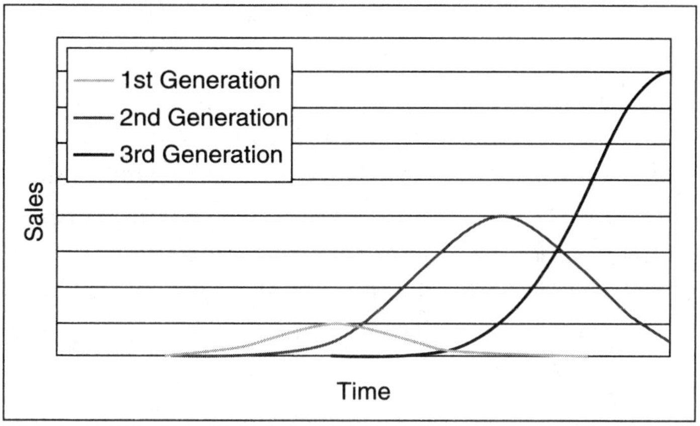

Figure 5-5 Technology lifecycles

As such, the forces of evolution tend to give rise to technology lifecycle curves that increase in size through successive generations, as depicted in Figure 5-5.[5]

> Technology lifecycle curves tend to increase in size through successive generations.

> In my analysis of the Hardware–Software Game, I assume that market potential (technology value) is a key driver of technology system adoption, and I assume technology markets[6] increase in size both over time and over successive generations.

5.3.2 Sensitivity to Price

It is incontrovertible to claim that price is generally a key driver of technology system adoption. I have discussed several underlying forces contributing to trends in user sensitivity to price over time.[7] One trend involves adoption by new users. I discussed in

5. Figure 5-5 was taken from Norton and Bass 1987.

6. The specific markets I refer to here are the "Served Available Markets" from strategic marketing literature.

7. See, for example, Bayus 1992; Danaher et al. 2001; Parker and Neelamegham 1997; and Parker 1992.

Chapter 2 the fact that as technology adoption passes through the different types of users, from innovators to early adopters to late adopters and finally laggards, users become increasingly sensitive to price.

Another trend involves validation and repeat purchases. As a technology system moves through the various stages of its lifecycle, users become comfortable with the performance and value of the system. As the risk and uncertainty surrounding purchase and usage of the system decrease, so does user sensitivity to price.

A third trend discussed in the technology lifecycle section involves production lifecycles. More specifically, as production passes through the early stages of system adoption and has proven itself in the market, associated profit opportunities attract more suppliers. Competition to provide more and different alternatives to the original system offerings leads to increases in the availability of system alternatives. The availability of more alternatives, in turn, increases user sensitivity to the price of any particular option.

Over time, user sensitivity to price

- Increases as adoption passes from earlier adopters to later adopters (first-time buyers)
- Decreases as uncertainty decreases (repeat purchasers)
- Increases as competition in supply increases

In my analysis of the Hardware–Software Game, I assume that the price of both hardware and software components is a key driver of technology adoption. In accordance with trends reported in the literature, (1) I assume that newer technology systems are relatively riskier than installed systems; (2) as such, I assume users are more sensitive to the prices of newer systems than to the prices of more established ones; and (3) I assume that the risks associated with any particular technology system, and thus user sensitivity to price, decrease over time.

5.3.3 Strength of Network Effects for Hardware

After market size and price, the third driver of technology system adoption is the strength of network effects for hardware. Direct network effects will be important in driving new system hardware adoption when the technology enables users to communicate or share resources with one another and/or early users provide validation of technology system value to later adopters. Indirect network effects will be important in driving new system hardware adoption when content creates a lot of value for users.

In my analysis of the Hardware–Software Game, I assume that network effects are similar from one generation of technology systems to the next; that is, I assume that if the presence of an installed base of users is important for driving sales of hardware for one generation of technology systems, then direct network effects will generally be important for driving hardware sales in other generations as well. Similarly, I assume that if the availability of content is important for driving sales of hardware for one generation of technology systems, then indirect network effects will generally be important for driving hardware sales in other generations as well.

Consistent with other research on network effects, I assume that the strength of network effects for hardware decreases over time.[8] I also test the impact of different relative strengths of direct and indirect network effects on the results of the game.

5.3.4 Strength of Network Effects for Software

The fourth driver of technology system adoption is the strength of network effects for software. A larger installed base of users will be important in driving demand for new technology software when having a lot of content and accessories enhances the value of the technology system to users. This effect mirrors the indirect network effects for hardware in the sense that if having a lot of content available attracts users to the system, then users will generally tend to buy a lot of content, which means software demand will be larger when there are more users.

A larger supply of available content will be important in driving new technology software demand when the users prefer a large variety of available content and/or when the tendency toward content saturation is small. Compare a PDA with a handheld gaming system. More content for the PDA generally means more and different applications that let you use your PDA for more and different things: longer-life batteries, word

8. See, for example, Bass 1969.

processor, GPS system, telephone, Internet access, and so on. More content for a portable gaming system means more games and movies. Content for PDAs, then, is likely to be complementary to other content, driving demand for ever more content, whereas content for portable game systems is likely to be substitutable for other content, eventually leading to saturation. At the same time, however, even if later users and/or content tend to complement earlier users and/or content, the market will eventually become saturated with users and/or content. As such, I assume that whether network effects start high or they start low, their strength decreases over time.

As with network effects for hardware, in my analysis of the Hardware–Software Game I assume that network effects for software are similar from one generation to the next; that is, I assume that if the presence of an installed base of users and/or the availability of content is important for driving software sales for one generation, then it will generally be important for other generations as well. As with network effects for hardware, I also assume that the strength of network effects for software decreases over time. I test the impact of different relative strengths of direct and indirect network effects on the results of the game.

5.3.5 Degree of Content Complementarity

The last driver of technology system adoption is the degree of content complementarity. This is the degree to which having more content provides more value for the technology system. In other words, does later content complement or increase the value of earlier content? If so, then the degree of content complementarity will be high, or equivalently, the tendency toward content saturation will be low. Or does later content substitute for, compete with, or otherwise decrease the value of earlier content? In this case, the degree of content complementarity will be low, or equivalently, the tendency toward content saturation will be high.

The example I used earlier of a system with a high degree of content complementarity is the bar-code scanner system. Applications for a bar-code system (credit card processing, inventory management, and point-of-sale data analysis) tend to complement one another. In contrast, applications for systems composed of hardware and consumables (copy/fax machines and paper/toner, hammers and nails, paintbrushes and paint, automobiles and gas/maintenance) tend to substitute for one another.

Similarly to the case of the strength of indirect network effects, regarding content complementarity, even if later content tends to complement earlier content, the market will eventually become saturated with content. As such, I assume that whether the degree of content complementarity starts high or it starts low, it decreases over time.

As with the magnitude of direct and indirect network effects, I assume that if content has a high degree of complementarity for one generation of a technology system, then it will have a high degree of complementarity for other system generations as well. I also assume that, whether it starts out high or low, the degree of content complementarity decreases over time.

5.4 Overview of the Game

Let's return to the timeline of events associated with the Hardware–Software Game as the adoption of a next-generation technology plays out in the marketplace.

At the time a next-generation technology is introduced into the marketplace (time zero), everyone is using an existing technology system. At this time, then, the market share of the old technology system is 100 percent, and the share of the new technology system is zero. As early adopters switch over to a new system, the installed base of users for the new system begins to grow, while the installed base of users for the old system begins to shrink. After early adopters have begun to adopt the new system, the corresponding network effects kick in and critical mass is eventually achieved, or adoption falters and the new technology system disappears from the market. Assuming the new system does take hold, it is around this time that adoption becomes self-reinforcing; that is, the value of switching to the new system increases as ever more users and content are attracted to the new system, and the value of staying with the old system falls as ever more users and content providers leave the old network.

Clearly, the most vital segment of the timeline occurs at the beginning, when adoption is starting to gain momentum and the new technology system either goes on to achieve critical mass, or it collapses; that is, parties with an interest in getting the new system adopted will face the greatest strategic challenge in trying to get users to upgrade during the early stages of adoption, as the base of new system users is established and moves toward critical mass. The challenge is to make the attractiveness to users of the new system relative to that of the old system as great as possible during this early stage of adoption.

Two questions are thus paramount:

1. For any given scenario, what will be the portion of users who upgrade to the new system during the earlier periods and what will be the portion of content allocated to the new technology during the earlier periods?
2. Which factors cause the transition to the new technology system to occur sooner rather than later?

I used my model of the Hardware–Software Game to simulate various scenarios of the game. The results are consistent with what we see in technology markets: Upon introduction of the new technology, hardware manufacturers have difficulties profitably selling hardware units to early adopters, given the lack of installed base and availability of content. More specifically, without having either existing users or a sufficiently large amount of content available, hardware manufacturers have to set unprofitably low prices to encourage users to adopt early on.

During the early stages of introduction of the new technology, content providers are serving the large installed base of old system users. They will not switch resources over to the production of new system content unless or until those resources will generate greater profits when used to produce new, rather than old, system content. Investment costs required to produce new technology content, together with network effects, induce greater economies associated with catering to a larger installed base of old system users. As a result, content providers are slow to switch resources over to the production of new system content, until a sufficiently large installed base of new system users has been established.

So in the early stages of new technology introduction the results of the interaction between independent parties, each of which acts in its own self-interest, is this: Hardware manufacturers are forced to subsidize sales to early adopters (that is, sell at a price below cost), and content providers supply a small amount of new system content at best, while continuing to dedicate the lion's share of resources to the production of content for the larger base of installed users of the old system. In other words, there is a chicken-and-egg problem. Hardware manufacturers cannot profitably sell a large number of units of new system hardware because there is not enough content (or users) to make the new technology valuable to users. At the same time, content providers will not supply content for the new system, because there are not enough new system users to buy it. Even if both hardware manufacturers and content providers would prefer rapid, large-scale adoption of the new system to slow adoption, the incentives do not exist for the two players, acting independently in their own self-interest, to take the actions necessary to achieve rapid adoption.

As the network of early adopters slowly grows and provides validation for the new technology system, the early majority may slowly adopt. If a sufficiently sized base of new technology users develops, it might eventually become profitable for content providers to move resources into the production of new technology content. If so, then the early majority will continue to adopt, eventually leading the new technology to achieve critical mass. At that point adoption will take off and the new technology will become established. This slow trickle of adoption toward critical mass is how the game has played out in the U.S. market for HDTV. The other possibility is that the number of adopters never reaches the point at which content providers find it more profitable to switch resources out of serving the old system market and into the new one. In the case of forestalled availability of new system content, adoption of the new system will lose momentum, and critical mass will not be achieved.

In Chapter 4, I described how the hardware manufacturers' problem is an inter-temporal one. Hardware manufacturers must figure out the best way to distribute sales of hardware units across periods so as to maximize total profits. In contrast, the content providers' problem is an intra-temporal one. Content providers must figure out the best way to distribute content each period across old and new system users so as to maximize total profits.

So, hardware manufacturers can be thought of as pushing users into the new system, trying to create forward momentum. In contrast, content providers are pulled back to the old system, drawn by the strong force of network effects. Hardware manufacturers want software providers to help them push users forward, yet content providers are drawn backward by the force of the old system. Generally, the initial phases of this tug-of-war favor inertia, that is, lack of forward movement into the new system. Only if hardware manufacturers can overcome this strong pull from the old system will they be able to achieve the momentum they need to achieve critical mass.

My simulation model helps shed light on the nature of the particular forces at work in the tug-of-war taking place under each configuration of network effects and technology characteristics. Having these detailed insights will help me figure out how best to weaken the forces pulling against and/or strengthen the forces pushing for each type of new system. The next section describes the various system configurations and characteristics I simulate with my model.

5.5 Simulation Categories and Scenarios

The three sets of players in the Hardware–Software Game—users, hardware manufacturers, and software providers—will have different incentives to take different actions for different technology systems that have different characteristics. If I want to understand how the Hardware–Software Game plays out generally, then, I must examine the structure of the game for various types of technology systems.

Different technology systems exhibit different levels of network effects. For example, the value users get from fax machines depends fundamentally on users' ability to interact with other users of fax machines. In contrast, users of PDAs can generate a lot a value by themselves, without having to interact with other PDA users. It follows that the strength of direct network effects for fax machines is much greater than the strength of direct network effects for PDAs. I should therefore expect the incentives and actions of the players in the fax adoption game to be different from those of players in the PDA adoption game.

At the same time, a particular technology system will exhibit smaller or larger network effects over the course of its technology adoption lifecycle; that is, when few people have fax machines (early adopters), new adopters of the technology system (early majority) will increase the benefits to new and existing owners dramatically. As more users adopt, however, the impact of the user base on increasing the value to new adopters and thus drawing in new fax machine users will diminish. Later, after most people have already bought fax machines, the addition of new owners (laggards) will add only small benefits to the existing base of users. In other words, the strength of network effects decreases over the adoption lifecycle of any particular technology system. I should therefore expect the incentives and actions of the players in the fax adoption game to be different during the early stages of the game from those during the later stages of the game.

I would like to know how the results of the Hardware–Software Game differ between technology systems like fax machines and PDAs. And I would also like to know how the results of the Hardware–Software Game play out during different stages of the adoption of particular systems, like fax machines. As such, I used my model of the Hardware–Software Game to simulate outcomes across different types of technology systems (four categories of network effects), as well as over the adoption lifecycle of particular types of systems (32 market scenarios).

I divide the types of systems I examine into four broad categories of network effects. For these four categories, differences between low network effects (L) and high

> I divide the types of systems I examine into four broad categories of network effects: LL, LH, HL, and HH. For these four categories, differences between low network effects (L) and high network effects (H) are relatively substantial, to account for large differences in the size of network effects across different types of systems.

network effects (H) are relatively substantial, *to account for large differences in the size of network effects across different types of systems*, such as fax machines and PDAs.

1. DNE, INE = L, L (Category LL)

   I classify technology systems as being in Category LL when they exhibit low (L) direct network effects (DNE) and low (L) indirect network effects (INE). These are systems in which users do not interact much with each other, and for which users' generation of value from using the system does not require a lot of content or accessories. Examples of systems with low direct network effects and low indirect network effects include hand tools (hammers and nails, screwdrivers and screws) and many kitchen appliances. With these two types of systems, users do not generally interact with one another or share resources, hence the low level of direct network effects. It is true that tools and appliances often need batteries or electricity to work, they sometimes require repair and maintenance, and users often benefit from having a variety of attachments, nails, bits, and screw sizes. These are all considered to be what I have been referring to as *software, content, and accessories*. However, relative to the value users generate from other systems by having more content available, the value users generate from tools and appliances does not increase *appreciably* with the availability of content, as compared, for example, with audio or video systems. More specifically, if you have a hammer, you can do a lot with unlimited quantities of one size of nail. However, with a CD player, you do not get a lot of value from having unlimited quantities of a particular CD.

2. DNE, INE = L, H (Category LH)

   I classify technology systems as being in Category LH when they exhibit low (L) direct network effects (DNE) and high (H) indirect network effects (INE). These are systems in which users do not interact much with each other or share resources, but for which users' generation of value requires a lot of content or accessories. Examples of systems with low direct network effects and high indirect network effects include PDAs, single-player gaming systems, TVs, CD and DVD players, and most other systems for which hardware acts as a conduit for accessing content. More specifically, users of *one-way systems* do not usually interact with other users, so direct networks effects are low. Yet having a larger base of users attracts a greater amount of content, which does provide value to users, so indirect network effects are high.

3. DNE, INE = H, L (Category HL)

   I classify technology systems as being in Category HL when they exhibit high (H) direct network effects (DNE) and low (L) indirect network effects (INE). These are systems in which users must interact with each other to generate value, but for which users' generation of value does not require a lot of content or accessories. Examples of systems with high direct network effects and low indirect network effects include fax machines, ATMs, and paintball. As in the case of users of systems with low direct network effects and low indirect network effects, users of systems with high direct network effects but low indirect network effects may very well benefit from having a variety of content, such as different transmission speeds for fax machines, added services for different ATMs, or different color paint for paintball systems. However, the value users generate from fax machines, ATMs, and paintball systems does not increase *appreciably* with the availability of content; that is, most of the value users get derives from the basic hardware and content offerings.

4. DNE, INE = H, H (Category HH)

   I classify technology systems as being in Category HH when they exhibit high (H) direct network effects (DNE) and high (H) indirect network effects (INE). These are systems in which users must interact with each other to generate value, and for which users' generation of value requires a lot of content or accessories. Examples of systems with high direct network effects and high indirect network effects include multiplayer gaming systems, cell phones, computer systems, transportation systems (vehicles and routes), and trading exchanges (NYSE, AMEX, NASDAQ, eBay, and so on). More specifically, users of *two-way systems* interact with other users, so direct networks effects are high. At the same time having a larger base of users attracts a greater amount of content, which also provides value to users, so indirect network effects are also high.

The four broad categories of network effects, LL, LH, HL, and HH, thus capture all the potential network effect configurations for different types of technology systems. Now I want to look at differences in conditions during the adoption lifecycle for each of the four types of technology systems. So for each of the four categories, LL, LH, HL, HH, I examine two different specifications for each of the five market drivers discussed earlier and summarized in Figure 5-6: (1) market size, (2) user sensitivity to price, (3) magnitude of direct network effects, (4) magnitude of indirect network effects, and (5) degree of software complementarity. Two possibilities for each of the five drivers create 32 (= 2 × 2 × 2 × 2 × 2) different market and technology scenarios for each category of network effects, as presented in Figure 5-6.

- Column [A] sets the identification number for each scenario configuration.
- Column [B] indicates whether market size is relatively low/small (L) or high/large (H).
- Column [C] indicates whether users are relatively sensitive (H) or insensitive (L) to price; that is, small changes in price lead to, respectively, large/high (H) or small/low (L) changes in demand for technology system hardware and software.
- Column [D] indicates whether the size of the installed base is more important (H) or less important (L) in creating value for the technology system. In the market scenarios, the difference between low (L) and high (H) direct network effects is relatively small, *to account for small differences during the different stages* of lifecycle adoption within a system. This contrasts with the difference between high and low direct network effects in the four categories of network effects just discussed. In the four categories of network effects, the difference between high and low direct network effects is relatively large to account for large differences in the size of network effects across different types of technology systems.
- Column [E] indicates whether software is more important (H) or less important (L) in creating value for the technology system. As with column [D], the difference between low (L) and high (H) indirect network effects for the market scenarios are relatively small, *to account for small differences during the different stages of lifecycle adoption within a* system.
- Column [F] indicates the degree to which more content complements (H) or substitutes for (L) earlier content.
- Column [G] indicates the relative profitability (1 = low, 2 = medium, or 3 = high) to hardware manufacturers of the technology system with the characteristics captured by columns [B] through [F].

- Column [H] indicates the relative profitability (1 = low, 2 = medium, or 3 = high) to software providers of the technology system with the characteristics captured by columns [B] through [F].

Scenario [A]	Market Size [B]	Sensitivity of: Users to Price [C]	Sensitivity of: HW Demand to Installed Base (DNE) [D]	Sensitivity of: HW Demand to SW = SW Demand to Installed Base (INE) [E]	Sensitivity of: SW Demand to SW Availability (SW Complementarity) [F]	Profit Index 1=L, 3=H: HW Profit Index [G]	Profit Index 1=L, 3=H: SW Profit Index [H]
1	L	L	L	L	L	2	1
2	H	L	L	L	L	2	2
3	L	H	L	L	L	1	1
4	H	H	L	L	L	1	1
5	L	L	L	H	L	2	2
6	H	L	L	H	L	2	2
7	L	H	L	H	L	1	1
8	H	H	L	H	L	1	1
9	L	L	H	L	L	3	2
10	H	L	H	L	L	3	2
11	L	H	H	L	L	2	1
12	H	H	H	L	L	2	1
13	L	L	H	H	L	3	2
14	H	L	H	H	L	3	3
15	L	H	H	H	L	2	1
16	H	H	H	H	L	2	1
17	L	L	L	L	H	2	2
18	H	L	L	L	H	2	2
19	L	H	L	L	H	1	1
20	H	H	L	L	H	1	1
21	L	L	L	H	H	2	2
22	H	L	L	H	H	2	3
23	L	H	L	H	H	1	1
24	H	H	L	H	H	1	1
25	L	L	H	L	H	3	2
26	H	L	H	L	H	3	3
27	L	H	H	L	H	2	1
28	H	H	H	L	H	2	1
29	L	L	H	H	H	3	3
30	H	L	H	H	H	3	3
31	L	H	H	H	H	2	1
32	H	H	H	H	H	2	1

Figure 5-6 Market/technology scenarios

5.6 Profit Frontiers by Category of Network Effects

Let's start with some definitions and clarifications. I define a *period* to be the length of time during which prices are fixed and cannot be changed by providers, due to transaction costs or whatnot. So all users who buy hardware or software during the same period pay the same price. Actual periods can be as little as one second—for example, the prices of stocks change almost continuously—or as long as several years, like the price of postage stamps, which must pass through regulatory processes before prices can be changed.

I define a *period* to be the length of time during which prices are fixed and cannot be changed by providers.

Since I am primarily interested in how the Hardware–Software Game plays out during the earlier stages of the game, I focus my simulation on the first two periods of the game. Hardware manufacturers (HW) introduce their new technology systems into the marketplace during the first period. At that time there is an installed base of users for the older technology system, and these users must decide whether or not to upgrade from the currently installed system to the new system. Users who have not yet adopted the older system may also choose to leapfrog the older system and jump right into the newer system. Also during the first period, content providers (SW) must decide how much of their resources to dedicate to the production of content for the currently installed system and how much to dedicate to production for the new system.

I define *profits* to be revenues less manufacturing costs. In particular, my definition of profits excludes any research and development costs incurred by HW to bring the new technology system to market. My definition of profits also excludes any sales, marketing, or administrative costs associated with technology system sales.

As I have discussed, the portion of users who transition to the new system during the first period will affect the market environment and potential profitability for HW and SW during the second period. How much, exactly, that effect will be will depend upon the particular market/technology scenario.

Before I move on to the profit surfaces, let me first reflect generally on how direct and indirect network effects affect HW's and SW's profit surfaces. Consider the case of cell phone systems, handsets, and services. Cell phone systems exhibit direct network effects, since cell phone users benefit from being able to communicate with other cell phone users. Cell phone systems also exhibit indirect network effects, since users benefit from having a multitude of products and services, such as greater coverage, the ability to send and receive text and voice messages, and the ability to use hands-free devices. What's more, users who communicate with one another (benefit from direct network effects) value being able to *both* call each other at certain times *and* send each other text messages at other times (benefit from indirect network effects). Cell phone users' ability to both call and text each other makes the ability of those users to communicate with each other even more valuable than being able to only text or only call one another; that is, as more content becomes available, direct network effects become larger, and when more people become members of the network, indirect network effects become larger. So the whole is greater than the sum of the parts.

What does this mean in terms of profits for HW and SW? Suppose there are two networks, one with one million members (Network 1M) and one with ten million members (Network 10M) and users can communicate only with people on their own network.

With positive direct network effects, Network 10M will generally provide more value to new adopters than Network 1M. This means that more potential users will want to join Network 10M than Network 1M (higher volumes for the HW of Network 10M than for the HW of Network 1M) and/or be willing to pay higher prices to join (higher margins for the HW of Network 10M than for the HW of Network 1M).

With positive indirect network effects, the availability of products and services creates value for users. Since Network 10M has more users, it will tend to provide greater profit potential, and thus tend to attract more providers of content and services than Network 1M. With greater content availability for Network 10M than for Network 1M, it follows that (1) more people will want to join Network 10M than Network 1M (higher volumes for the HW of Network 10M than for the HW of Network 1M) and/or be willing to pay higher prices to join (higher margins for the HW of Network 10M than for the HW of Network 1M), and (2) each cell phone user in Network 10M will want more services than users in Network 1M (higher volume per user for the SW of Network 10M than for the SW of Network 1M) and/or users will be willing to pay more for services (higher margins for the SW of Network 10M than for the SW of Network 1M).

So now, based on this discussion, it becomes apparent that both direct and indirect network effects affect HW's profits; that is, HW generate higher profits when user interaction is more important and more users adopt their systems, and HW also generate higher profits when content creates a lot of value and there is more content available. In particular, HW are able to extract some of the value created by the amount of con-

When there are positive *direct network effects,* larger networks will generate

- Higher sales volumes for HW
- Higher margins for HW

When there are positive *indirect network effects,* larger networks will generate

- Higher sales volumes for HW
- Higher margins for HW
- Higher sales volume per user for SW
- Higher margins for SW

tent made available by SW. Yet only indirect network effects affect SW; that is, SW's profits are higher when content creates more value and there is more content available. But direct network effects—that is, value created by user interaction—do not have a direct impact on SW's profits. Rather, direct network effects increase SW's profits only indirectly, by increasing the amount of content each user buys.

Think of it this way. A technology system is composed of two groups of components, the hardware components and the software components. For cell phone systems, the hardware is the handset, and the software is the connection services and features. The hardware components of the system tend to be durable, or long-lasting; that is, users buy the handset once. Some users may use the handset more or less intensely than others, but regardless of how much time users spend using the handset, they have to buy it only once.[9]

In contrast, the software components (connection services) of a system tend to be nondurable; that is, if users want to continue to utilize content over long periods of time (for the same length of time they use the hardware), they will tend to have to pay for software many times (users have to pay for connection services every month). In particular, people who spend more time using their phone tend to use more content. They go through more minutes each month and/or they use more different kinds of services, such as voice services, text services, and so on. So the decision users face for

9. Of course, users who use the hardware more intensely may end up using the hardware for less time, and then having to replace the hardware, relative to someone who uses the hardware less intensely. But these are the people who would be more likely to adopt the technology in the first place; that is, the need to buy hardware more often will generally not cause high-intensity users to not adopt the technology.

hardware is generally a one-time, yes-or-no decision. In contrast, the decision users face for content is an ongoing, how-much-do-I-use-each-period decision.

When potential users are deciding whether or not to buy the hardware (and how much they are willing to pay for it), they consider what the value of the hardware is (how many other users have adopted it), and they also consider how much software they will be able to access. If more people have adopted the system, then potential users will be more likely to adopt the system (buy the hardware) and/or pay a higher price to adopt the system. Similarly, if a lot of software is available, then potential users will be more likely to adopt the system and/or pay a higher price to adopt the system. Both direct and indirect network effects thus affect HW's profitability.

When system users are deciding how much content to buy, they consider the extent to which content creates value (the size of indirect network effects): How many minutes per month will I use? Will I use text services? Users also consider the extent to which content complements or substitutes for other content (the degree of content complementarity). If I have a lot of minutes to talk, do I still need to be able to send text messages? The answers to both sets of questions will likely depend on the number of people who are members of the network.

Some types of technology system users will tend to use more services when the size of the network is larger; that is, content per person will increase as the size of the network increases. Cell phone users will tend to use more content as the size of the network increases and they have more people with whom they can communicate. For other types of systems, though, the average user will tend to use less content as the size of the network increases. In the case of DVD players, for example, the first adopters are likely to be those people who love watching movies. In this case, early adopters will tend to watch a lot of movies; that is, content per user will be high when there are few users. In contrast, those people who wait to adopt DVD players until the price has come way down tend to be those people who watch fewer movies. In this case, content per new adopter will be low when the size of the network is large. If this pattern does, in fact, hold for DVD players, content per user will decrease as the number of DVD users increases.

More generally, when direct network effects are high, content per user will tend to increase as the size of the network increases. Contrarily, when direct network effects are low, content per user will tend to decrease as the size of the network increases. Notice, though, that this interaction between direct network effects and indirect network effects is actually captured through the specification of indirect network effects; that is, the interaction between network size and demand for content is captured by having content per user increase or decrease with the size of the network. In particular, the demand for software is affected only by the magnitude of indirect network effects (how much content users consume, and whether this amount changes with the size of the network). The magnitude of direct network effects, though, does not influence user demand for content directly. Rather, it is captured indirectly by having the magnitude of indirect network effects change as the size of the network changes.

So what we are left with is that both direct and indirect network effects affect the demand for hardware and thus HW's profits. In contrast, only indirect network effects, and not direct network effects, affect the demand for content, and thus SW's profits.

Both direct and indirect network effects affect the demand for hardware and thus HW's profits. In contrast, only indirect network effects, and not direct network effects, affect the demand for content, and thus SW's profits.

As an aside, a good example of content complementarity can be found in sequels—books, movies, games, and so on. When a sequel to, say, a movie comes out, users will be more likely to see the sequel if they have seen the original. For example, one of the reasons *Spiderman 2* was such a hit was that so many people had seen the original *Spiderman*. Many, if not most, of the people who saw *Spiderman* also saw *Spiderman 2*. Because of this sequel effect, Columbia Pictures knew they were guaranteed a minimum amount of revenues from theater, DVD, and accessory sales of *Spiderman 2*. It gets better, though. When *Spiderman 2* came out, it generated revenues for Columbia Pictures from sales of *Spiderman 2* products and accessories. In addition, however, the release of *Spiderman 2* also increased sales of the original *Spiderman* products and accessories, because people went back to see the original *Spiderman* again, either to prepare for or to extend the enjoyment of *Spiderman 2*. And when *Spiderman 3* came out, it increased sales of *Spiderman* and *Spiderman 2* products and accessories. This forward and backward spillover from one product to another is called a *halo effect*.

Now let's move on to the profit surfaces. Figures 5-7A through 5-7D illustrate sample potential profit curves for HW and SW. There is a separate set of graphs for each of the four categories of network effects, LL, LH, HL, and HH, as defined earlier. Each set of graphs contains four diagrams, one each for first- and second-period profits, each for HW and SW. Since the profit curves for the second period are dependent upon the level of adoption during the first period, for the sample illustrations provided, I assume that 10 percent of potential users adopt the new system during the first period.

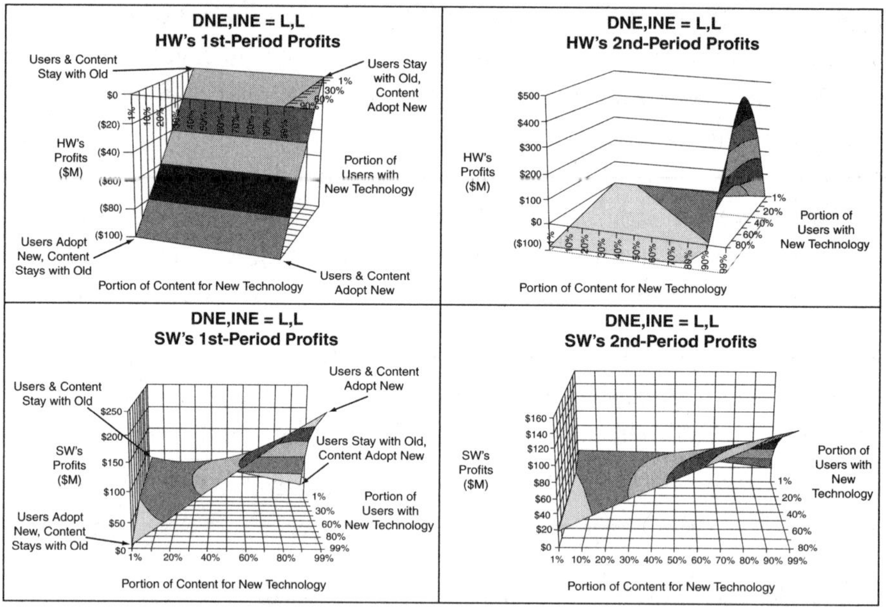

Figure 5-7A HW and SW profit surfaces: Category LL (DNE,INE = L,L)

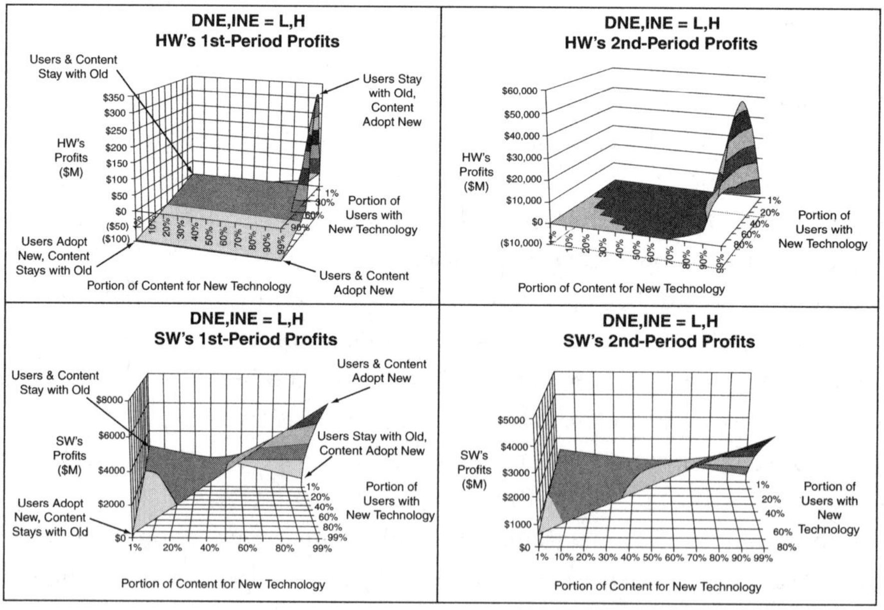

Figure 5-7B HW and SW profit surfaces: Category LH (DNE,INE = L,H)

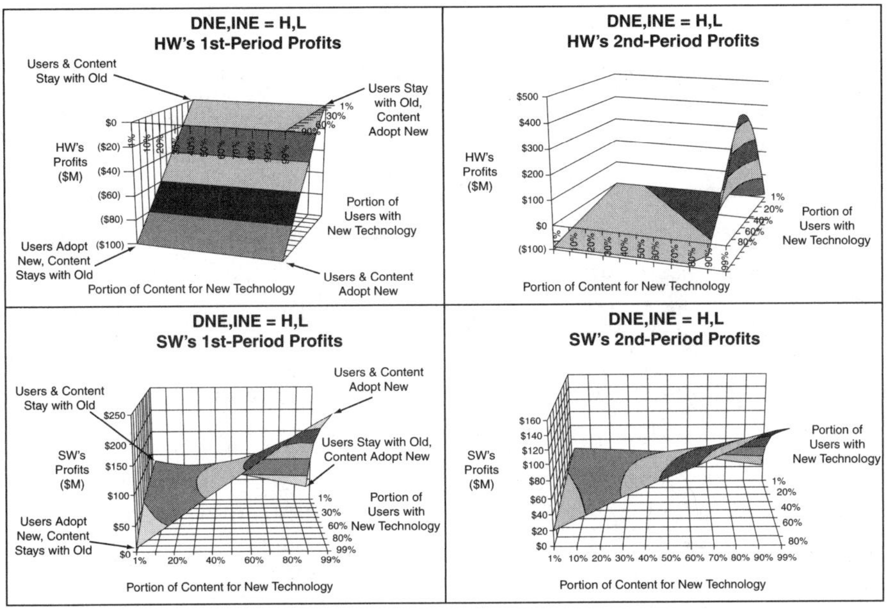

Figure 5-7C HW and SW profit surfaces: Category HL (DNE,INE = H,L)

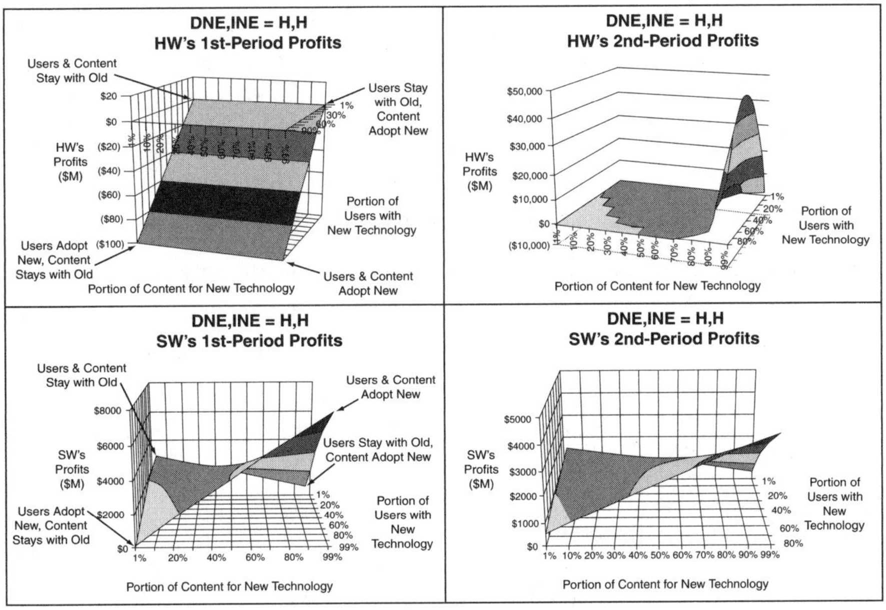

Figure 5-7D HW and SW profit surfaces: Category HH (DNE,INE = H,H)

5.6.1 Hardware Manufacturers' Profits

Consider the first set of graphs, Figure 5-7A, those for the network category in which both direct and indirect network effects are low (DNE, INE = LL). Let's say the old system is a screwdriver, chisel, or shovel and the new system is an electric drill. Both screwdrivers and drills can be used to make holes of different sizes in different types of surfaces and to insert and remove different types of screws and rivets. Direct network effects occur when a user of one type of screwdriver or drill can easily operate other types of screwdrivers or drills, or when a group of users can work together to build a house, clear a concrete road, or drill for oil more effectively than an individual. Indirect network effects occur when different types and sizes of holes need to be dug or screws need to be inserted. Content providers supply different sizes and types of screws that will work with both manual screwdrivers and electric drills. Content providers can also supply drill bits of various types and sizes. However, limits on resources mean that if content providers supply more types of drill bits, they will have to supply fewer types of screws.

The graph in the upper left-hand corner of Figure 5-7A shows the profits HW will earn during the first period when the new system (the electric drill) is introduced into the marketplace. HW's profits will depend on both (1) the portion of content available for the new system (drill bits for making different-size holes or inserting different kinds of screws) (SW's actions) and (2) the portion of potential users who adopt the new system (users' actions). The graph in the upper left-hand corner shows HW's level of profits for each combination of the other two players' sets of actions.

5.6.1.1 Hardware Manufacturers' First-Period Profit Surfaces

Since there is no installed base during this first period of introduction to create direct network effects, the value of the technology system to first-period adopters will reflect only the technology value (as defined in Chapter 2) and the amount of content available for the new system.

The technology value to users of an electric drill is the ability to more quickly and accurately drill holes or screw in screws than they could with a manual screwdriver or shovel. One person alone building a cabinet could construct the cabinet more quickly with an electric drill than with a screwdriver. Using an electric drill, one person alone on a construction site can work a bit more quickly than coworkers who are using manual screwdrivers. But as early adopters, those people would have to learn how to use the drill on their own, and unless many holes need to be drilled or screws need to be screwed in over and over again, the time savings of the electric drill over the manual one may not be substantial.

Let's make the following assumptions:

- The early cost of manufacturing an electric drill is $50. After some users have adopted, the cost drops to $25, due to learning by doing and economies of scale. The basic drill comes with a bit for working on wood, and the cost of a special drill bit for working in concrete, supplied by the content provider, is $10.
- A drill lasts for one working day, eight hours.
- Construction workers earn $80 an hour if they work just with wood, or $100 an hour if they can also work with concrete.
- Once they have learned how to use the drill, workers will end up saving a half-hour a day by using an electric drill, rather than a manual screwdriver, on wood. Assume further that workers will save an hour a day using the drill on concrete. So, workers can effectively work an extra half-hour a day on wood or an extra hour a day on concrete when they use an electric drill instead of a manual screwdriver.
- For early (first-day) adopters who must learn how to use the drill on their own, it will take the most adept an hour to learn how to use the drill; it will take the second most adept two hours to learn how to use the drill; it will take the third most adept three hours to learn how to use the drill; and so on. Later (second-day) adopters can benefit from experiences and help from early adopters. It will therefore take later adopters half the time to learn how to use electric drills as it will take early adopters.

If the most adept type of worker buys a drill during the first day to work on wood, that worker will lose an hour of work learning to use the drill and $80 in pay (see Figure 5-8, row [A1]), after which time the worker will end up making up half of the lost hour of time, earning $40 in extra wages (row [A2]), by using the drill rather than a screwdriver. A drill for working on wood will therefore end up costing the worker $40 in lost wages, relative to the case in which the screwdriver alone was used (row [A3]). In other words, for Worker 1 to be willing to adopt the electric drill, HW would have to give away the drill for free plus pay the worker $40.

If that worker buys a drill plus an extra bit to work on concrete, he or she will lose an hour of work, or $100 in pay, learning to use the drill (row [A5]) but will make up an hour working on concrete, earning $100 (row [A6]), leaving the worker even time-wise with or without the drill. However, adopting the drill will cost the worker $10 for

	Day 1 No Installed Base of Users [A]	Day 2 With Installed Base of Users [B]
Worker 1		
Wood		
[1] Time Cost	1.0 hour @ $80/hour = ($80)	0.5 hour @ $80/hour = ($40)
[2] Time Savings	0.5 hour @ $80/hour = $40	0.5 hour @ $80/hour = $40
[3] Value of Drill to Worker 1	($40)	$0
[4] HW Profit	$40 payment – $50 cost of drill = ($90)	$25 cost of drill = ($25)
Concrete		
[5] Time Cost	1.0 hour @ $100/hour = ($100)	0.5 hour @ $100/hour = ($50)
[6] Time Savings	1.0 hour @ $100/hour = $100	1.0 hour @ $100/hour = $100
[7] Cost of Concrete Bit	($10)	($10)
[8] Value of Drill & Bit to Worker 1	($10)	$40
[9] HW Profit	$10 payment – $50 cost of drill = ($60)	$40 price – $25 cost of drill = $15
Worker 2		
Wood		
[10] Time Cost	2.0 hours @ $80/hour = ($160)	1.0 hours @ $80/hour = ($80)
[11] Time Savings	0.5 hour @ $80/hour = $40	0.5 hour @ $80/hour = $40
[12] Value of Drill to Worker 2	($120)	($40)
[13] HW Profit	$120 payment – $50 cost of drill = ($170)	$40 payment – $25 cost of drill = ($65)
Concrete		
[14] Time Cost	2.0 hours @ $100/hour = ($200)	1.0 hours @ $100/hour = ($100)
[15] Time Savings	1.0 hour @ $100/hour = $100	1.0 hour @ $100/hour = $100
[16] Cost of Concrete Bit	($10)	($10)
[17] Value of Drill & Bit to Worker 2	($110)	($10)
[18] HW Profit	$110 payment – $50 cost of drill = ($160)	$10 payment – $25 cost of drill = ($35)

Figure 5.8 Electric drill example

the concrete drill bit (row [A7]). In other words, for the worker to adopt the drill, HW would have to give the worker the drill for free, plus pay an extra $10 to buy the concrete bit from SW (row [A8]).

We now have HW having to pay the worker $40 to adopt the drill for woodworking, leaving HW a net loss of $90, including the cost of the drill (row [A4]). If the worker uses the drill on concrete, HW will have to pay the worker $10 to adopt the drill and buy the concrete bit, leaving HW a net loss of $60 (row [A10]), assuming SW make the concrete drill bit available.

The second most adept type of worker will spend two hours learning to use the drill to save a half-hour of time on wood or an hour of time on concrete. Buying the drill to do woodwork will therefore cost that worker $160 in lost wages to generate $40 in extra wages, for a net loss of $120 (rows [A10]–[A12]). To convince the second most adept type of worker to adopt the drill, HW would therefore have to pay $120, leaving HW with a net loss of $170, including the cost of manufacturing the drill (row [A13]). Alternatively, buying the drill to do work on concrete will cost the worker $200 in lost wages to generate $100 in extra wages, less $10 for the cost of the drill bit, for a net cost of $110 (rows [A14]–[A17]). To convince the second most adept type of worker to adopt the drill, HW would therefore have to pay $120 to adopt the drill for woodworking or $110 for working on concrete (rows [A12], [A17]). So if the second type of worker buys the drill to work on concrete, the net loss to HW is $170 if the drill is used on wood or $160 if the drill is used on concrete (rows [A13], [A18]).

We are not quite done. HW must offer the same terms to all workers who buy drills during the same period. So, if both types of workers adopt the drill for work on wood, HW will end up with a net loss of $170 for each worker, for a total loss of $340. If, on the other hand, concrete drill bits are available, and both types of workers adopt the drill for use on concrete, HW will end up with a net loss of $160 for each worker, for a total loss of $320.

Now assume that HW cannot pay workers to adopt the drill, that the most HW can do is give the drill away for free. In this case, the lowest profit HW can earn when each worker adopts the new system is ($50), the cost of the drill. In this case, HW's total profit will be ($50) if one worker adopts, ($100) if two workers adopt, and so on.

There are several important points illustrated in the example:

- Worker 1 will get greater benefits from adopting the electric drill than Worker 2. So Worker 1 will adopt the drill earlier than (or no later than) Worker 2. In other words, users who value a new system the most will be the first to adopt, and users who value a new system the least will be the last to adopt.
- Since Worker 1 values the new system more than Worker 2, Worker 1 will be willing to pay a higher price to adopt the new system than Worker 2. However, HW must offer both workers the same terms if both of them adopt during the same day. To get both users to adopt during the same day, then, HW must charge all adopters the price that the user with the lowest value will pay. In other words, as more users adopt within the same period, the prices HW can charge each user fall. As more users adopt, HW thus exchange margin (higher profits per adopter) for volume (more adopters).

> During the first period of new technology introduction:
>
> - Users who value a new system the most will be the first to adopt.
> - As more users adopt, HW exchange margin (higher profits per adopter) for volume (more adopters).
> - Users are willing to pay higher prices for hardware when more content is available.

- Both worker types are better off—that is, they are willing to pay higher prices for hardware—when more content is available.

Return now to Figure 5-7A and consider HW's first-period profit surface for the case in which both direct and indirect network effects are low (DNE, INE = LL). The first thing to notice is that HW's first-period profits are everywhere negative. This means that for the particular parameters used in the example, the value of the new technology system to early adopters when there is no installed base is too low to enable HW to charge a profitable price during the first period (when manufacturing costs are still high). In other words, HW will generally charge prices to early adopters that are less than the costs of manufacturing early hardware units, as in the electric drill example just presented. Since prices will be less than costs, HW will generate losses during the first period on every unit sold. Also, since HW must lower the price to get more users to adopt, both per-unit and total losses will be larger when more users adopt early on. In other words, the surface representing HW's profits during the first period slopes downward (becomes more negative) as the portion of first-period adopters increases.

> HW will generally charge prices to early adopters that are less than the costs of manufacturing early hardware units, thus generating losses during the first period.

The second thing to notice is that HW's first-period per-user profits are the same, regardless of the portion of users who adopt or the portion of content available for the new system; that is, the profit surface is a linear plane, rather than becoming increasing negative (convex) as the portion of users who adopt increases and as the availability of content decreases. This is a by-product of a restriction I placed on the model. As I indicated in the previous paragraph, the negative profits for HW mean the prices users are willing to pay for new system hardware when there is no installed base are less than the cost of providing hardware, for all levels of content availability. In the model, I set the lowest possible price equal to zero. This means that in the worst case, HW can give the hardware away to users for free, but I assume that HW cannot pay users to adopt the new system. In this latter case of paying users to adopt, prices would be negative, rather than just zero, as in the drill bit example.

The fact that HW's profits are the same regardless of the level of content available is a result of this restriction that prices cannot be negative. It places an upper limit on the losses HW can generate from each unit sold, the loss limit being the cost of manufacture ($50 in the drill example). If I were to lift this restriction and allow HW to pay users to adopt (as in the drill example), then HW's first-period profits would vary with the level of adoption and with content availability, where the prices HW would have to pay each user to adopt would be less when fewer users adopt and when more content is available. In other words, in the profit surface in Figure 5-7A, the value of the system to users is, in fact, greater when there is more content available. However, regardless of content availability, the value to users of system hardware is still less than its cost of manufacture. Said differently, with the restriction of non-negative prices, the value to users under the most favorable conditions (content availability of 100 percent) is still too low to convince users to pay a positive price.

5.6.1.2 Hardware Manufacturers' Second-Period Profit Surfaces

In the electric drill example, I assume that during the second day there is an installed base of users who bought the drill and learned how to use it during the first day. During the second day, help from the installed base of users means that it will take new adopters only half the time to learn how to use the drill as it would had they adopted during the first day. Also, since HW now have more experience in manufacturing drills, it costs only $25 to make them, half of what it cost to make them the previous day.

Working through the example during the second day, if Worker 1 adopts the electric drill for woodworking, the time spent learning to use the drill exactly offsets the time saved. So if Worker 1 adopts the drill to work on wood, he or she will be willing to pay $0, leaving HW a profit net of manufacturing costs of ($25) (see Figure 5-8, rows [B1]–[B4]). If concrete bits are available and Worker 1 adopts the drill to work on concrete,

HW can charge that worker a price of $40 (rows [B5]–[B8]). With the cost of manufacturing the drill, HW will thus earn a profit of $15 (row [B9]). If Worker 2 adopts the drill to work on wood, HW must pay that worker $40, so, including manufacturing costs, HW will incur a net loss on the sale of $65 (row [B14]). And if Worker 2 adopts the drill to work on concrete, HW must pay that worker $10, so HW will incur a net loss on the sale of $35 (row [B20]).

During the second day, then, if Worker 1 adopts the drill, HW's profit will be ($25) when no extra drill bits are available, and $15 if SW do make drill bits available. If another worker adopts, remembering that HW must offer both types of workers the same terms, then HW's profit *per adopter* drops to ($65) when no bits are available and ($35) per adopter when bits for working on concrete are available.

Important points illustrated by day 2 of the example include the following:

- As on the first day, as workers start to adopt during the second day, HW's profits per adopter rise. As more workers adopt, and HW must offer lower prices to all adopters, profits per adopter fall. So, as more users adopt, HW exchanges margin (higher profits per adopter) for volume (more adopters).
- As on the first day, both workers are better off, and therefore willing to pay higher prices for hardware, when more content is available.
- During the second day, there is an installed base of users who adopted the new system during the previous day. Users who adopt the second day value the system more, and are therefore willing to pay higher prices, when there is an installed base of users.

Now consider HW's second-period profit surface in Figure 5-7A, that is, the graph in the upper right-hand corner. This graph shows the level of profits HW will earn for

> During the second period of new technology introduction:
>
> - As more users adopt, HW exchanges margin (higher profits per adopter) for volume (more adopters).
> - Users are willing to pay higher prices for hardware when more content is available.
> - Users are willing to pay higher prices for hardware when there is an installed base of users.

various levels of second-period (1) content availability and (2) user adoption, assuming 10 percent of users adopt the new system during the first period; that is, the profit curve reflects the fact that there is an installed base of users (10 percent of total potential users), which creates value through direct network effects for second-period adopters. If I were to assume that even more users adopted the technology during the second period, then the impact of direct network effects would be greater. In this case, second-period users would be willing to pay even higher prices to adopt the new system than those shown, so that the profit surface would shift upward (see Figures 5-10A and 5-10B).

Notice that HW do not earn positive profits unless content availability is very high. The large area of unprofitability indicates that in this example the value of the system to new adopters is too low for them to adopt when there is too little content available (in the electric drill example, even with the availability of the concrete bit, Worker 2 would still not be willing to pay a positive price to adopt the electric drill). Equivalently, the value of the system to new users will be high enough to enable HW to charge new adopters profitably high prices only when there is a sufficient amount of content available for the new system (about 80 percent or more of the total software available for the situation depicted in the graph). As such, the profit curve becomes higher to the right as more content becomes available. This reflects the fact that with more content availability for the new system, HW will be able to charge higher prices and earn higher profits from sales of hardware to second-period adopters.

Finally, there is an inverted-U-like shape for HW's profits when content availability is high. This reflects the fact that as more users adopt (volume increases), HW's profit per adopter falls (margin decreases). HW's total profits are the product of margin and volume, as seen in Figure 5-9, where **Profits per Adopter** is based on the price that the adopter with the lowest value for the new system is willing to pay.

Starting from zero, the initial adopters—that is, those who value the technology the most—will be willing to pay high prices, so HW will generate large profits (high margins) for each of these initial adopters (low volume). At this point of low volume, HW can usually increase total profits by lowering the price a bit to attract a few additional adopters. What HW loses in margins (lower prices) will be more than offset by

HW's Total Profits = [Profits per Adopter] x [Number of Adopters]

Figure 5-9

greater numbers of adopters (higher volumes). So during the period, as HW moves from no adoption through some adoption, total profits will rise.

As the process continues, HW will reach a point at which profits per adopter and number of adopters are both moderate. At this point, if HW lower prices further to attract additional adopters, more in profits will be lost from the current adopters, who are willing to pay a higher price, than will be gained from the additional adopters; that is, after a certain point, when prices are low enough and volume is high enough, further sacrifices in profits will outweigh any additional gains in volume, and total profits will fall.

This trade-off of margin for volume is represented by the concave (first increasing then decreasing) shape of the profit curve on the right-hand side of HW's second-period profit surface in Figure 5-7A. When almost all content is made available for the new system, as the portion of new adopters during the second period rises from zero to about 50 percent, HW's total profits increase; that is, the added profit from each additional user more than compensates for the loss in profit from users who would have been willing to pay a higher price. However, after HW reaches the price at which half of users are willing to adopt, efforts to further increase adoption by lowering prices will end up generating more losses than gains. In other words, the graph says that when all content is made available for the new system, HW will maximize second-period profits by setting the price so that half of potential users adopt.

Now compare HW's first-period profit surfaces across technology systems with different combinations of direct and indirect network effects. HW's first-period profit surfaces have the same shape for three of the four combinations of network effects, LL, HL, and HH. This suggests that the lack of an installed base for new technology systems and their associated direct network effects will generally lead HW to earn negative profits during the first period. This is not so surprising, given that I am looking at precisely those technology systems for which direct network effects are important in creating value for system users. When there are too few users to generate much value through direct network effects, systems are not valuable enough to new users for them to be willing to adopt at prices that are high enough to cover manufacturing costs. The exception is Category LH. When direct network effects are low and indirect network effects are high, having an installed base of users is not that important for creating system value to users, but content is. For this category of technology systems, HW will be able to generate positive profits early in the adoption process, if there is a large enough amount of content available.

Finally, compare HW's second-period profit surfaces across technology systems with different combinations of direct and indirect network effects. In all four cases,

> HW will earn positive profits only if there is a lot of content available.

LL, LH, HL, and HH, HW will earn positive profits only if there is a lot of content available, and in all of these cases, HW will do best by setting a price so that half of potential users adopt each period.

Also, HW's second-period potential profits are larger when direct network effects are higher. This makes sense. Systems with larger network effects have the potential to create more value for users, and HW can generate more profits from users when there is more value created from which to extract the profits.

> HW's second-period potential profits are larger when direct network effects are higher.

Last, for all but the highest levels of content availability, HW's second-period profit surfaces (1) are relatively flat when indirect network effects are high (LH and HH): HW's second-period profits hover around zero until the portion of content for the new technology system becomes very high; (2) tilt downward when indirect network effects are low (LL and HL): HW's second-period profits become increasing negative, as the portion of content for the new technology system becomes low and the portion of new adopters becomes high. To get a better idea of what is happening here, I examined the impact of first-period adoption on HW's second-period profit surfaces. Figure 5-10A shows HW's second-period profit surfaces for differing levels of first-period adoption when indirect network effects are low, and Figure 5-10B shows profit surfaces when indirect network effects are high. Figures 5-10A and B illustrate how HW's second-period profit surfaces change as the portion of first-period adopters increases, and with it, the value of direct network effects for second-period adopters. The series of figures can also be interpreted as the evolution of HW's profit surfaces over successive periods as more users adopt. What the figures illustrate is that unless the new technology system provides enough value to users, HW will be unable to charge a profitably high price at which potential users will adopt. More specifically, when *both* direct *and* indirect network effects are low, the new system will not provide enough value for the masses to adopt at profitable prices, unless or until there is *both* a large enough installed

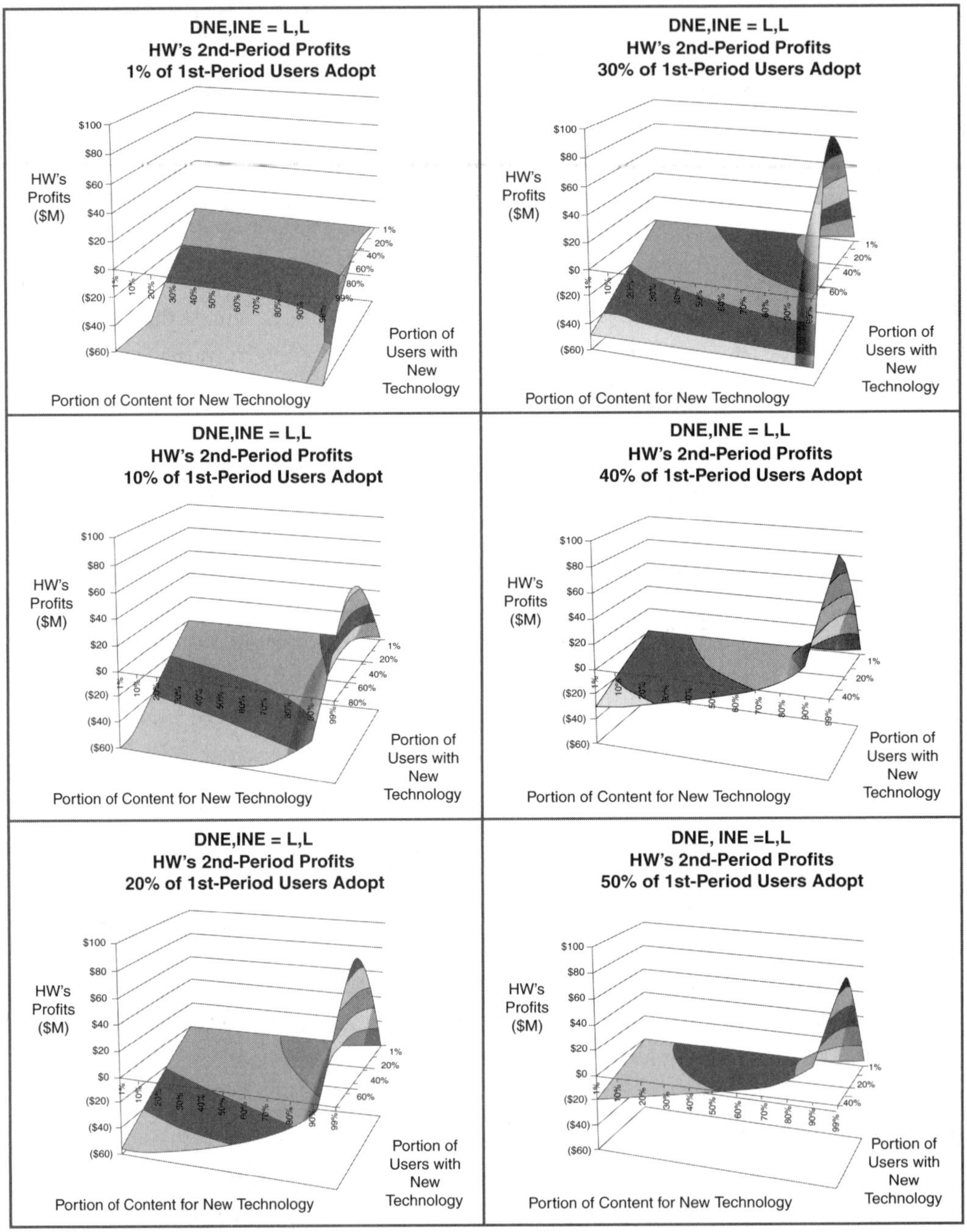

Figure 5-10A Hardware manufacturers' second-period profits under alternative sizes of the installed base: Category LL (DNE,INE = L,L)

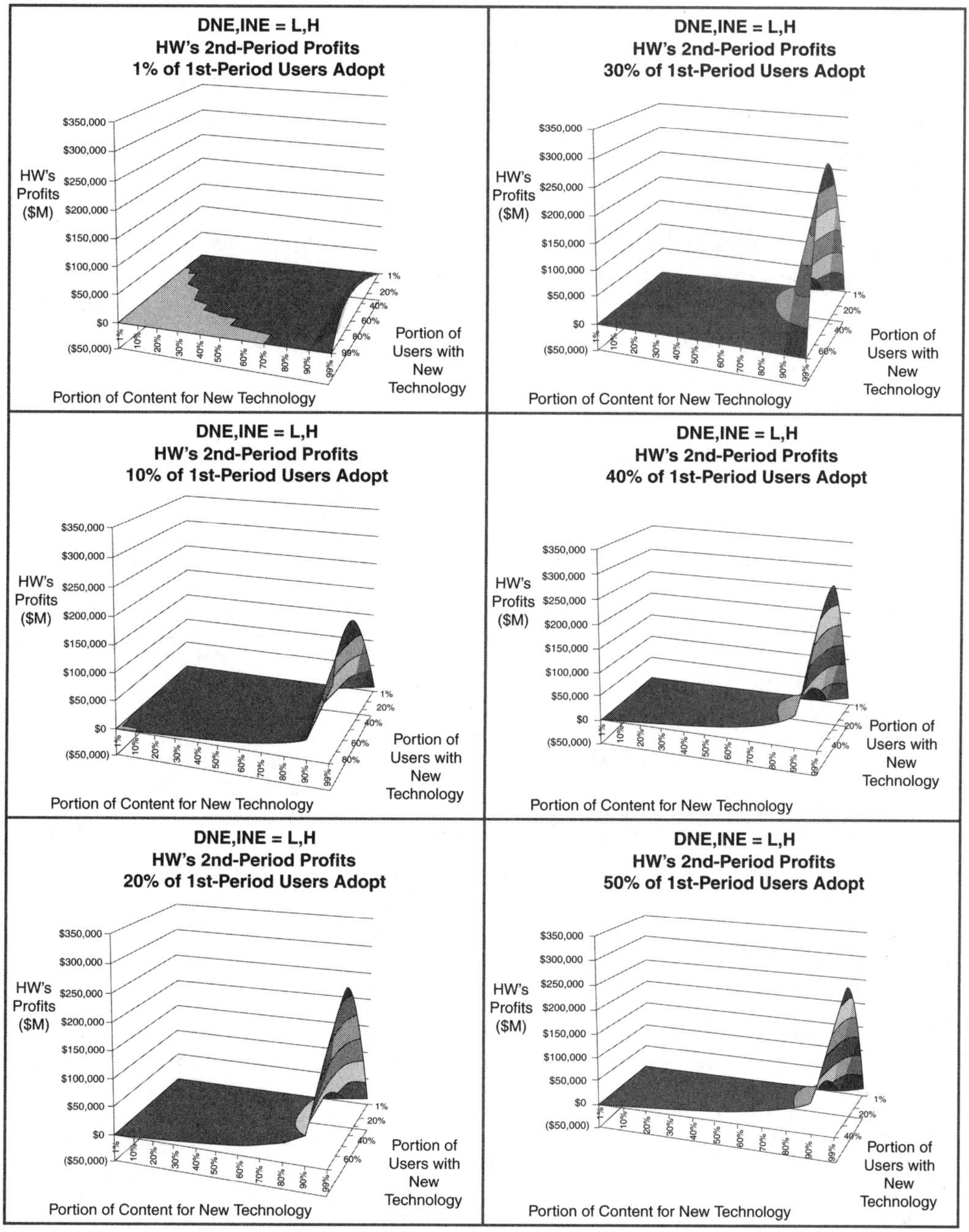

Figure 5-10B Hardware manufacturers' second-period profits under alternative sizes of the installed base: Category LH (DNE,INE = L,H)

> When *both* direct *and* indirect network effects are low, the new system will not provide enough value for the masses to adopt at profitable prices, unless or until there is *both* a large enough installed base *and* enough content available.
>
> When *either* direct *or* indirect network effects are high, the new system will provide enough value to the masses, as long as there is *either* a large enough installed base *or* enough content available.

base *and* enough content available. On the other hand, when *either* direct *or* indirect network effects are high, the new system will provide enough value to the masses, as long as there is *either* a large enough installed base *or* enough content available.

5.6.1.3 Summary of Hardware Manufacturers' Profits

To summarize, HW's profit surfaces share the following characteristics:

There must be enough content available for HW to be able to generate positive profits. When there is not enough content available, the value of new technology systems to users is too low for them to be willing to pay high enough prices for HW to cover costs. When technology systems exhibit both direct and indirect network effects, but direct network effects are small, software plays a relatively more important role in creating value for users. In these cases, HW can profit from introducing new technology systems only when there will be a large amount of content available. Conversely, when direct network effects are large, the availability of content is less important for creating a profitable environment in which to introduce new technology systems. In these cases, HW will need less content to be available to make it worthwhile to bring new systems to market.

HW's profits are larger when there is more content available for new systems. When technology systems exhibit indirect network effects, having more content available creates greater system value for system users. What's more, even though SW supply content, HW are able to extract part of the value created for users by the availability of content in the form of higher hardware prices and thus higher profits.

HW will do best by leading users to adopt new systems over several periods. There is a trade-off to HW between higher profits per new adopter and more new adopters each period. HW will realize greater profits when users adopt slowly than when they adopt all at once. More specifically, HW will optimize profits across all periods by setting prices so as to encourage half of the remaining potential users to adopt the new technology each period. This is the classic monopoly result, which means that monopolists in network industries will have the incentives to behave the same as monopoly suppliers for more traditional industries.

5.6.2 Software Providers' Profits

Let's now consider SW's profit surfaces, the two graphs in the bottom half of Figures 5-7A through 5-7D. As with HW, SW's profits will depend on both (1) the portion of content available for the new system (SW's actions) and (2) the portion of potential users who adopt the new system (users' actions). The surface graphs in the lower left-hand corner of Figures 5-7A through 5-7D show SW's level of profits for each combination of these two sets of actions during the first period. At this time, the new technology system has just been introduced into the marketplace, and there is no installed base of users for it. The surface graphs in the lower right-hand corners show SW's level of profits for each combination of SW's and users' actions during the second period, after some users have adopted the new system and created direct network effects. As I mentioned above, in the examples presented in Figures 5-7A through 5-7D of SW's second-period profit surfaces, I assume 10 percent of users adopted the new system during the first period.

5.6.2.1 General Overview of Software Providers' Profit Surfaces

The Hardware–Software Game from SW's standpoint provides an almost perfect parallel to the Prisoners' Dilemma Game (PDG), as described in section 5.1. In the PDG, the more "adverse" action is to rat out one's partner, and the more "favorable" action is to keep quiet. Similarly, from SW's profit standpoint, the more "adverse" action is to stay with the old system, and the more "favorable" action is to adopt the new system. In the PDG, Larry and Linda are best off if both keep quiet, they are better off if they rat each other out, and they are worst off if one keeps quiet while the other rats. Similarly, from SW's perspective, SW and users are best off if both adopt the new technology system, they are better off if both decide to stay with the old system, and they are worst off if one party adopts the new system while the other party stays with the old system.

In SW's profit surfaces in Figures 5-7A through 5-7D, the back left-hand corner of the graphs shows profits to SW when both SW and users choose to stay with the old system. The upper right-hand corner of the graphs shows SW's profits when both SW and users choose to adopt the new system. The two remaining corners, the lower front left-hand corner and the lower back right-hand corner, show SW's profits when one party chooses to stay with the old system while the other party adopts the new system. I define the *diagonal* of the profit surface to be the path that is formed as both users and SW *move together* from no adoption to full adoption; that is, the diagonal is the path that starts with no adoption of the new system by either set of players in the lower left-hand back corner of the profit surface. The path continues as both sets of players move together through some adoption of the new system and through more adoption of the new system. And the path ends when both sets of players have fully adopted the new system, at the upper right-hand corner of the profit surface.

> I define the *diagonal* of the profit surface to be the path that is formed as both users and SW *move together* from no adoption to full adoption.

Suppose the new technology system is only slightly better than the old system. This is akin to saying that both Larry and Linda will get only a bit more prison time if they rat each other out than they will get if both keep quiet. In this case, the diagonal in SW's profit surface will be relatively flat; that is, the height of the profit surface at the point of full adoption (upper right-hand corner) will be only slightly higher than the height of the profit surface at the point of no adoption (lower left-hand corner). At the other extreme, if the new system provides much more value than the old system, the diagonal in SW's profit curve will have a large slope; that is, SW's profits when both sets of players adopt the new system will be much greater than their profits when both sets of players stay with the old system.

The off-diagonal corners of SW's profit surface are those regions where the available supply of content does not match the base of users; that is, in the lower left-hand portion of the graph, most users have adopted the new system, but most content is provided for the old system. Likewise, in the lower right-hand portion of the graph,

> The greater the value of the new system is relative to the old one, the steeper the slope of the diagonal will be.

most users are staying with the old system, but most of the content is provided for the new system. In both of these cases there is a mismatch between the players' actions. And in both of these cases, SW will be worse off (will earn lower profits) than when both sets of parties (SW and users) agree to stay with the old system or, better yet, move together to the new system.

5.6.2.2 Software Providers' First-Period Profit Surfaces

The most interesting aspect of SW's first-period profit surfaces lies in the path of the diagonal. As the first few users and the first bit of content start to move together from the old system to the new one, SW's total old and new system profits decrease. In fact, as users and content start to move together from the old system to the new one, SW are worse off until a critical mass of users and content has been established for the new system. This is reflected in the profit surfaces in Figures 5-8A through 5-8D by the convex nature of the surfaces, that is, by the fact that profits are lower in the center of the surfaces than they are at either end of the diagonal (the lower-left and upper-right corners of the graphs).

> As users and content start to move together from the old system to the new one, SW are worse off until a critical mass of users and content has been established for the new system.

This phenomenon is seen more clearly in Figure 5-11. The figure illustrates the path of SW's relative profits as users and content start to migrate to the new technology system. The figure indicates that when indirect network effects are larger, more users must transition to the new system before SW are better off than when all users and content were with the old system. (Recall from the discussion in section 5.6 that the magnitude of direct network effects does not affect SW's profits directly. Rather, the impact of direct network effects on SW's profits is captured by the changing nature of content per user as the size of the network changes, that is, through the size of indirect network effects.)

To reiterate, SW do better by having the bulk of their content *either* for the old system *or* for the new system. SW are actually worse off when just a few users adopt the new system than when all users stay with the old system, even if the new system will ultimately generate higher profits for SW after all users and content have switched.

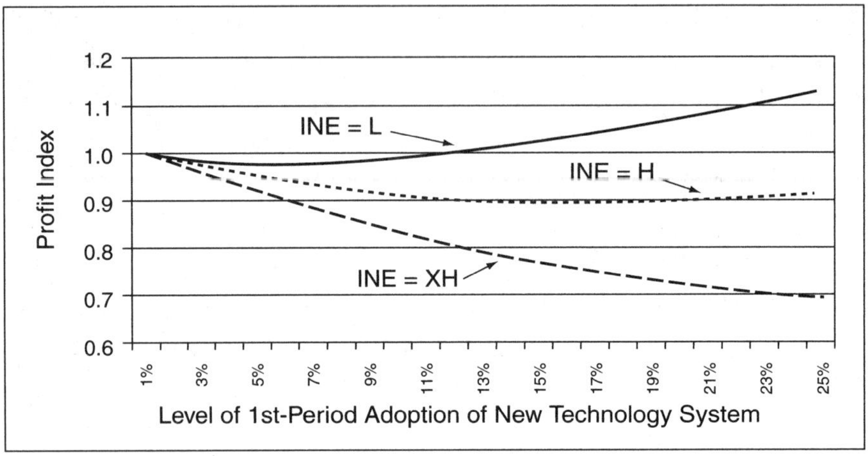

Figure 5-11 Content providers' first-period profits under alternative levels of first-period adoption

This convexity is due to the "multiplier effect" associated with networks, where the value of the total is greater than the sum of the parts. This idea was discussed in more detail in Chapter 1, and it was illustrated in Figure 1-3. There are more potential links, and thus generally more potential value, when all users are together in one network, as opposed to being divided among multiple, smaller networks. This multiplier effect associated with networks means it is generally better to have all content in one place, *either* with the old system *or* with the new system, rather than dividing it among different groups. This convexity associated with strong indirect network effects will have a deterrent effect on SW's incentives for switching from the provision of older system content to newer system content. SW will be reluctant to produce much content for the new system until enough users have switched to the new system. The portion of users who must switch before SW will want to provide content for the new system will be larger when indirect network effects are stronger.

5.6.2.3 Software Providers' Second-Period Profit Surfaces

Now consider SW's second-period profit surfaces in Figures 5-7A through 5-7D. SW's second-period profit surfaces have the same general shapes as do the first-period surfaces, but the second-period surfaces tend to be somewhat flatter. This is a by-product of my assumption that the magnitude of network effects decreases over time. If I were, instead, to assume that the strength of network effects remained constant over time, then the second-period profit surfaces would be as curved as the first-period surfaces.

SW's second-period profit surfaces share the same phenomenon as the first-period surfaces in that as users and content begin to move together from the old system to the new one, SW's profits decrease. As in the first period, SW are worse off during the second period when just a few users adopt the new system than when all users stay with the old system, and their losses are larger as the strength of direct network effects increases. SW will start to be better off with the new system only when a substantial portion of users have transitioned. Network effects thus deter SW from transitioning from the old to the new system.

We have established that network effects dampen SW's incentives to transition from the old system to the new one during any particular period. But how do network effects impact SW's adoption momentum over time? In other words, if users have started to adopt the new system during previous periods, will SW resist or support further adoption? Figure 5-12 provides a series of three graphs illustrating SW's second-period profits. The first graph shows SW's second-period profits for technology systems in which indirect network effects are low (INE = L); the second graph, systems in which indirect network effects are high (INE = H); and the third graph, systems in which indirect network effects are extra-high (INE = XH). Each graph shows SW's second-period profits for combinations of different levels of first- and second-period adoption of the new technology system.

What the series of graphs in Figure 5-12 shows is that SW will tend to resist transition to the new technology system more when indirect network effects are larger. This makes sense. When indirect network effects are large, the large installed base of users for the old system creates more profit potential for SW than the small base of new system adopters. Furthermore, the base of installed users for the old system will continue to create more value for SW until "enough" users have switched to the new system. When network effects are stronger, more users will have to switch before it pays for SW to also switch. This means SW are more likely to resist transitioning to the new system (SW have more to lose by transitioning) until more users have transitioned. This resistance by SW to transitioning to the new system when network effects are higher means that once momentum for the transition to the new system begins, it will take much longer and move much more slowly for systems with stronger indirect network effects than for systems with weaker network effects.

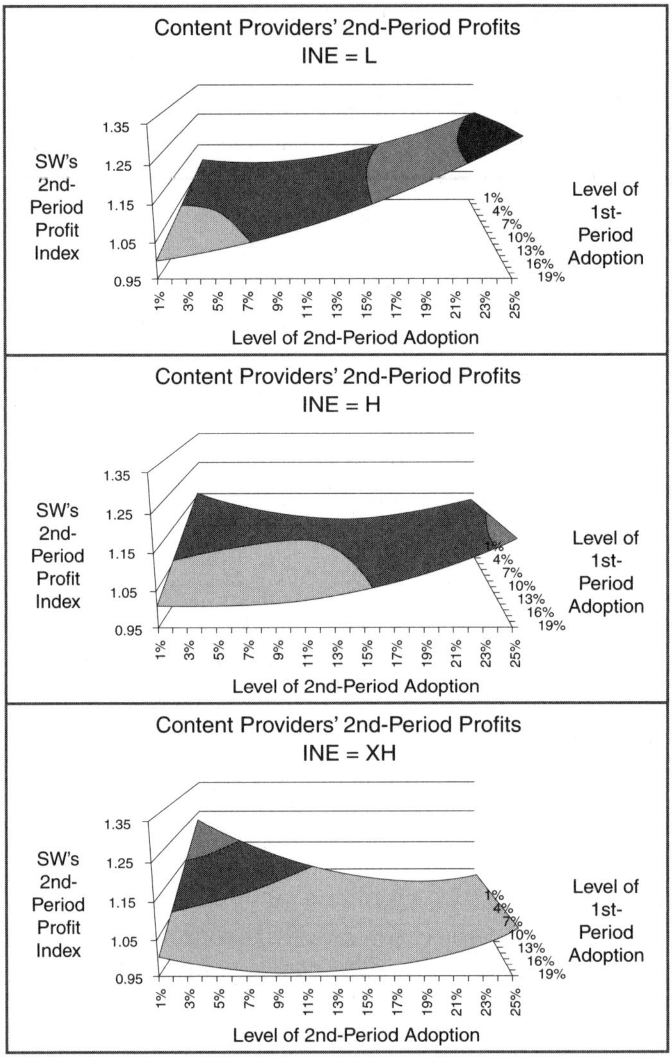

Figure 5-12 Content providers' second-period profits under alternative levels of adoption

> Resistance by SW to transition to the new system will be greater when network effects are higher.

5.6.2.4 Summary of Software Providers' Profits

To summarize, SW's profit surfaces share the following characteristics. Assuming that, relative to the old technology system, the new system provides greater benefits for users and greater profit opportunities for SW when all parties have transitioned, the points of highest profits for SW in descending order of profitability are:

- Both users and SW fully adopt the new technology system.
- Both users and SW fully remain with the old technology system.
- One party remains with the old system while the other party adopts the new system.

SW are actually worse off when just a few users adopt the new system than when all users stay with the old system, even if the new system will ultimately generate higher profits for SW after all users and content have switched. This convexity associated with strong indirect network effects will have a deterrent effect on SW's incentives for switching from the provision of older system content to newer system content. In fact, SW will be reluctant to produce much content, if any, for the new system until enough users have switched to the new system. The threshold portion of users who must switch before SW will want to provide content for the new system will be larger when indirect network effects are stronger. Finally, this resistance by SW to transition to the new system when network effects are higher means that once momentum for the transition to the new system begins, it will take much longer and move much more slowly for systems with stronger indirect network effects than it will for systems with weaker network effects.

5.7 Impact of Speed of Adoption on Profitability

After gaining a sense of what the profit trade-offs to hardware and software providers look like for each of the two periods, I next looked at how the speed of adoption affects overall profits for hardware and software providers across the two periods. For each of the 32 market scenarios defined in Figure 5-6, I used the simulation model to calculate the profits hardware manufacturers and content providers would earn using five different adoption rate specifications, increasing from slow to fast rates of adoption.

In the first scenario (1,1/90,90), that with the slowest rate of adoption, I assume that during the first period, 1 percent of content is provided for the new technology system and 1 percent of users adopt the new system. I calculated first-period profits under this assumption of 1 percent content and user adoption. For the second period, I assume 90 percent of content is provided for the new technology system and 90 percent of users adopt the new technology system. I calculated second-period profits under this assumption of 1 percent existing base of users from the first period and 90 percent content and user adoption during the second period. I then aggregated first- and second-period profits assuming a 10 percent rate of discount.

The second scenario (10,10/80,80) provides a faster pace of adoption than the first scenario. In the second scenario I assume 10 percent of users and content transition to the new technology system during the first period and 80 percent transition during the second period.

In the third scenario (20,20/70,70) I assume 20 percent of users and content transition to the new technology system during the first period, and 70 percent transition during the second period.

In the fourth scenario (30,30/60,60) I assume 30 percent of users and content transition to the new technology system during the first period, and 60 percent transition during the second period.

The fifth and last scenario (40,40/50,50) I examined provides the fastest pace of adoption of the five scenarios. In the fifth scenario I assume 40 percent of users and content transition to the new technology system during the first period, and 50 percent transition during the second period.

The paces of first-period adoption that yield the highest profits for hardware manufacturers versus software providers are displayed in Figure 5-13. The comparisons of HW's and SW's profitability under different speeds of adoption confirm that the two parties prefer very different rates of adoption under every scenario and network effect configuration. However, it is not always HW that prefer the speediest pace of adoption.

For HW, the optimal speed of adoption of new technology systems will depend on the extent to which earlier adopters create value for later adopters, that is, the size of direct network effects. When direct network effects are weak, early system adopters do not create a lot of value for later adopters. In this case, hardware manufacturers are less apt to sacrifice profits in the short run—that is, subsidize adoption by a large number of users early on—since this sacrifice will not create large benefits in later periods. As a result, when direct network effects are weak, HW will tend to prefer slower rates of adoption.

Conversely, when direct network effects are strong, HW *do* have an incentive to encourage more users to adopt during earlier periods, since they will create a lot of

> When direct network effects are weak, HW will tend to prefer slower rates of adoption. When direct network effects are strong, HW will tend to prefer faster rates of adoption, but not too fast.

value for later adopters, and thus greater potential profits for HW. So when direct network effects are strong, HW will tend to prefer faster rates of adoption.

At the same time, however, even when direct network effects are high, and the prospects of greater value for later adopters push HW to encourage more early adopters, HW do not want the speed of adoption to be too fast. This is because getting more users to adopt means lower prices; that is, profits per new adopter will be lower (or losses per adopter will be higher) when there are a whole lot of new adopters than when there are fewer new adopters. This tendency to want to generate profits per adopter that are as high as possible will tend to moderate the optimal pace of adoption for HW, even when direct network effects are strong. The exception is when direct network effects are strong and indirect network effects are weak. In this case profits per adopter for early adopters will be relatively low regardless, so HW do not sacrifice much in profits per user by encouraging mass adoption earlier. These effects are seen in Figure 5-13.

For technology systems with stronger direct network effects (DNE = H), the optimal pace of adoption for HW is at least as fast as, and in most cases faster than, the

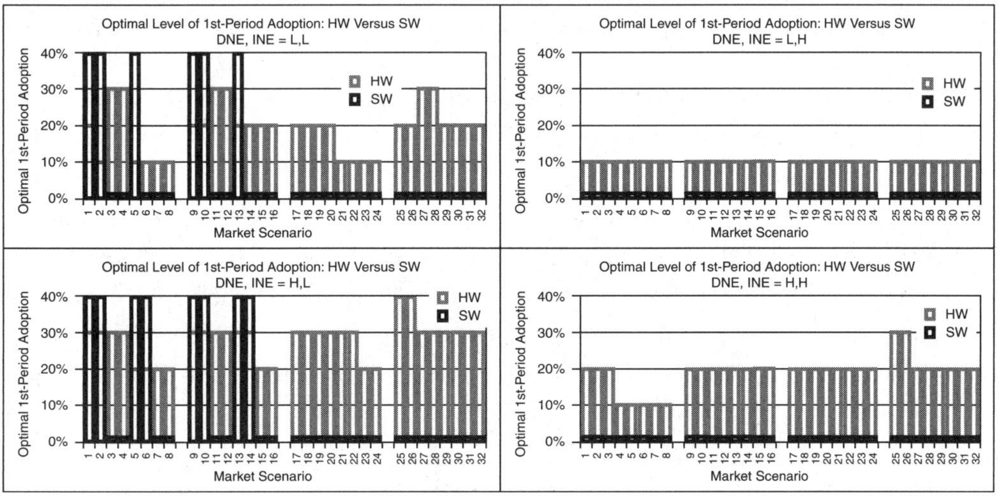

Figure 5-13 Optimal level of first-period adoption: HW versus SW

optimal pace of adoption when direct network effects are weaker (DNE = L). Furthermore, during the early stages of adoption of particular systems when direct network effects tend to be stronger (Market Scenarios 9–16 and 25–32), HW prefer speedier rates of adoption than in the later stages when direct network effects tend to abate (Market Scenarios 1–8 and 17–24). Finally, note that hardware manufacturers tend to prefer moderate rates of first-period adoption, 10 to 30 percent, rather than more extreme rates of adoption, 1 percent or 40 percent.

As for SW, the preferred speed of adoption is dependent upon the profitability of the older installed technology system versus the (potential) profitability of the newly introduced technology system. If serving old system users is more profitable than serving new system users, SW will want to dedicate all their resources to providing content for old system users, and vice versa. In other words, SW will prefer a very slow pace of new system adoption when serving old system users is profitable. Conversely, SW will prefer a very rapid pace of adoption when the new system provides better opportunities than the old one. These effects are seen in Figure 5-13, where SW always prefer either very slow rates of adoption (1 percent) or very fast rates of adoption (40 percent).

SW will prefer a very slow pace of new system adoption when serving old system users is profitable. Conversely, SW will prefer a very rapid pace of adoption when the new system provides better opportunities than the old one.

Generally speaking, when direct network effects are low, the value associated with the technology system will lie in the system itself, and in the availability of content when indirect network effects are strong. In this case, since I assume the new system is inherently more valuable than the old one, software providers will prefer to move to the new system as quickly as possible. In the opposite case, when direct network effects are strong, a technology system is of more value when the installed base of users is large. In this case, during the early stages of adoption of the new system, the large installed base of users is (currently) that for the older system. So software providers will prefer to serve that large base of old system users; that is, they will prefer slow adoption of the new system.

5.8 Sensitivity of Profits to Changes in Market Drivers

I have looked at how the players' various potential actions (when to adopt the new system, what price to charge for hardware, and how much content to supply) affect their payoffs. I have also looked at how the speed of adoption of the new system affects the players' payoffs. The last question I asked myself was this: Of the various market drivers I have been considering (market size, content complementarity, user sensitivity to price, and magnitude of direct and indirect network effects), which are the most important in affecting the players' payoffs? In other words, for which of the drivers will small changes in magnitude cause large changes in HW's or SW's potential profits?

Of course, at this point it should be clear that larger market sizes, larger network sizes, and stronger network effects provide greater potential profit opportunities for technology system suppliers. But what I am really interested in learning is the changes in which of the market drivers will provide the largest bang for my buck; that is, for which of the market drivers will relatively small changes in the size of the driver lead to large changes in potential profits? Digging down deeper, the real issue here is this: For which of the drivers will small changes in the size of the drivers lead to large changes (1) in the value users derive from the system and/or (2) in the portion of the value created for users that HW and SW can extract?

To answer these questions I calculated the changes in HW's and SW's profits as the sizes of the various market drivers change. For each of the four network categories (LL, LH, HL, HH), I calculated HW's and SW's two-period profits for each of the 32 market/technology scenarios under the fastest adoption rate specification (40,40/50,50). To see how profits changed as the size of each driver changed, I compared these two-period profits for the cases in which each market driver was small/low (L) to those in which the market driver was large/high (H), while holding all other factors constant. In other words, separately for HW and SW, and separately for each category of network effect (LL, LH, HL, and HH), I made 16 different comparisons. For example, in the case of market size, I compared profits for Market Scenario 1 to those for Market Scenario 2, profits for Market Scenario 3 to those for Market Scenario 4, profit for 5 to those for 6, 7 to 8, 9 to 10, 11 to 12, 13 to 14, 15 to 16, 17 to 18, 19 to 20, 21 to 22, 23 to 24, 25 to 26, 27 to 28, 29 to 30, and 31 to 32, separately for HW and SW, and separately for each of the four network configurations.

The results of these profit comparisons are summarized in Figures 5-14 and 5-15. The numbers reported are elasticities, that is, the percentage change in profits when the market drivers increase by 1 percent. Larger elasticities mean bigger impacts on profits for the same change in the sizes of the market drivers. For example, an impact on profits equal to 0.5 means that a 1 percent change in the size of the market driver will increase profits by half a percent. Similarly, an impact on profits equal to five means that a 1 percent increase in the size of the driver will increase profits by 5 percent. More generally, small increases in the sizes of market drivers will increase profits by only a small amount when elasticities are less than one, but by a large amount when elasticities are greater than one.

When both indirect network effects are low and the degree to which later software complements earlier software is low (row [5.1] of Figure 5-14), the environment does not provide much opportunity for HW or SW to generate profits. In these cases, small increases in the market drivers tend to cause profits to become less negative, to switch from negative to positive, or to grow from very low levels to moderate levels. In many of these cases, either percentage changes in profits are not well defined, or they can be very large. As such, these figures do not tend to provide useful insights into the dynamics of the game. These figures are reported in Figure 5-14 but not illustrated in Figure 5-15, and I exclude them from further discussions.

	Hardware Manufacturers								Content Providers							
	Low User Sensitivity to Price				High User Sensitivity to Price			Low User Sensitivity to Price				High User Sensitivity to Price			
Market Driver [A]	DNE = L [B]		DNE = H [C]		DNE = L [D]		DNE = H [E]	DNE = L [F]		DNE = H [G]		DNE = L [H]		DNE = H [I]	Impact on Profits [J]
[1] Market Size	0.2 – 0.3		0.2 – 0.4		0.1 – 0.2		0.1 – 0.5	0.2		0.2		0.1 – 0.2		0.1 – 0.2	Low
[2] Sensitivity to Price					(2.0)		(2.0)					(2.0) – (1.7)		(2.0) – (1.7)	
[3] Content Complementarity				>							>				Medium
[3.1] INE Low	6 – 72	<	10 – 36		9	>	9 – 33	8 – 14	<	13 – 50		4 – 17	<	6 – 8	
[3.2] INE High	6	<	6		2 – 4	<>	3	7	=	7		3 – 4	=	3 – 4	
[4] Direct Network Effects				<											
INE Low															
[4.1] SW Comp Low	3 – 4	<	4 – 16		N/A		7 – 24	0.0		0.0		0.0		0.0	
[4.2] SW Comp High	3 – 5	<	4 – 5		4 – 7	>	3 – 13	0.0		0.0		0.0		0.0	
[4.3] INE High	2 – 3	<	4		2 – 3	<	3	0.0		0.0		0.0		0.0	
[5] Indirect Network Effects				>							>				High
INE Low															
[5.1] SW Comp Low	65	>	43 – 56		N/A		N/A	40 – 55	<	49 – 116		60		N/A	
[5.2] SW Comp High	37 – 54	>	35 – 37		N/A		25 – 55	31 – 32	=	31 – 32		15 – 18	=	15 – 18	
[5.3] INE High	59 – 60	>	58 – 59		23 – 34	>	23 – 25	52	=	52		20 – 21	=	20 – 21	

Notes:
Figures are elasticities, that is, the percentage change in profits with a 1 percent change in the size of the market driver.
N/A: Elasticities are not well defined.

Figure 5-14 Sensitivity of profits to changes in market drivers

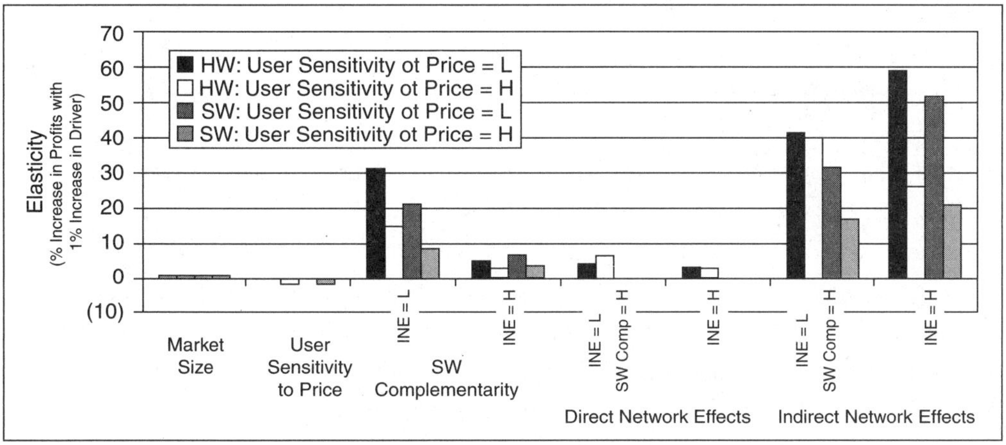

Figure 5-15 Increase of profits with increase in size of market driver

The first thing to notice about the analysis presented in Figures 5-14 and 5-15 is that profit opportunities for both HW and SW increase by only a small amount when market size increases (section [1] in Figure 5-14) or user sensitivity to price decreases (section [2] in Figure 5-14), when all other market drivers are held constant. Profit opportunities for HW and SW increase moderately when the degree of software complementarity (section [3] in Figure 5-14) or the magnitude of direct network effects (section [4] in Figure 5-14) increases. And profit opportunities for HW and SW increase substantially when the magnitude of indirect network effects increases (section [5] in Figure 5-14). This reflects the idea that the most value for users is created when indirect network effects are large. Likewise, increases in the degree of software complementarity or the magnitude of direct network effects create moderate amounts of value for users.

The second thing to notice about the analysis is that profit opportunities tend to be larger when users are less sensitive to price. This is seen in the table by comparing column [B] to column [D], column [C] to column [E], column [F] to column [H] and

> The most value for users is created when indirect network effects are large.
>
> Increases in the degree of software complementarity or the magnitude of direct network effects create moderate amounts of value for users.

> Profit opportunities tend to be larger when users are less sensitive to price.

column [G] to column [I]. In these cases, users are willing to pay for value; that is, HW and SW can extract the value created for users in the form of higher prices.

In accordance with the discussion above, profit opportunities for HW and SW will be the greatest when the technology system creates a lot of value for users, and *at the same time*, HW and SW are able to extract this value. This occurs when indirect network effects are higher and at the same time users are less sensitive to price.

Finally, the ability of increases in the degree of content complementarity or the magnitude of direct network effects to increase potential profits is larger when indirect network effects are low. In other words, there are decreasing returns to increasing amounts of content complementarity or direct network effects. At the same time, the ability of increases in the magnitude of indirect network effects to increase potential profits is larger when indirect network effects are high; that is, there are increasing returns to increasing amounts of indirect network effects.

> There are decreasing returns to increasing amounts of content complementarity or direct network effects.
>
> There are increasing returns to increasing amounts of indirect network effects.

Why are profits more sensitive to the size of indirect network effects than to the size of direct network effects or the degree of content complementarity?

Consider CD technology systems. Users of CD systems can interact with one another by sharing CDs, but generally speaking, users tend to generate more value by using CD players by themselves, rather than by interacting with other users. CD systems thus exhibit relatively small direct network effects. In contrast to inter-user links, content plays a large role in creating value for CD system users. The more types of CDs there are available, the more value users get. To the extent that CD systems have more applications—that is, users can burn their own CDs as well as play prerecorded DVDs containing music, books, lectures, and so on—then having more content will increase the value of CD systems to users even more. So, CD systems exhibit large indirect network effects, and content complementarity (being able to use blank CDs, music CDs, book CDs, and so forth) further enhances the strength of indirect network effects.

Contrast CD systems with cell phone systems. Users of cell phones can interact with one another by calling or texting other cell phone users, and in fact, the primary value users get from cell phones is being able to interact with other users. Cell phone systems thus exhibit large direct network effects. At the same time, users can generate value by using their cell phones by themselves, by personalizing ring tones, by creating picture galleries, and by accessing the Internet. The availability of more and different kinds of content (voice and text services, Internet access, ring tones, and so on) enhances the value to users of using their cell phones *both* while interacting with other users *as well as* while using their phones themselves. Cell phone systems thus exhibit large indirect network effects.

More generally, technology system users get value from using the systems by themselves (listening to CDs or creating cell phone picture galleries), and they also get value from interacting with other system users (sharing CDs or calling other cell phone users). The users' total value from being members of the system is thus the sum of (1) the value they get from using the system themselves and (2) the value they get from each link to other system users, times the total number of links, as displayed in Figure 5-16.

Stronger direct network effects mean users get more value from being connected to other system users. So when direct network effects are stronger (as with cell phones), the value of being linked to every other member of the network (the size of **B** in Figure 5-16) increases.

Stronger indirect network effects mean content enhances the value to users of using the system by themselves and/or content enhances interaction between system users. So when indirect network effects are stronger (as with both CD systems and cell phones), the availability of content will increase the value to users of using the system by themselves (having more CDs or ring tones to choose from; that is, the size of **A** in Figure 5-16 is larger) and/or the value of being linked to other system users (being able to exchange CDs or text or call other cell phone users; that is, the size of **B** in Figure 5-16 is larger).

System Value to Users = (Value by Themselves) + [(Value per Link) x (# Links)]

V = A + [B x C]

Figure 5-16

When software is more complementary (music and books and lectures, or text and voice and Internet services), greater availability of software has the ability to increase system value to users either more quickly or to a greater extent. In either case, when software is more complementary, the terms **A** and **B** in the relationship in Figure 5-16 will either be larger or grow faster than when software is less complementary.

Now we see that when direct network effects are stronger, the value per link (term **B**) will be larger. When indirect network effects are stronger, both the value of the technology itself (term **A**) and the value of being connected to other system users (term **B**) will be larger. Since indirect network effects have the ability to enhance the values both of intra-action and of inter-action, while direct network effects have the ability to enhance the value only of inter-action, it follows that larger indirect network effects will have the ability to increase the profit opportunities of system suppliers more so than larger direct network effects.

A final note involves profit extraction by HW. If the indirect network effects associated with a technology system are strong, then HW can potentially earn large profits from sales of hardware. However, for HW to be able to profit from the value created for users by the indirect network effects, HW must be able to actually extract this value.

Take, for example, the recently resolved standards war between manufacturers of Blu-ray and HD-DVD over next-generation DVD technology systems. Before the battle was resolved, suppliers were reluctant to make many Blu-ray or HD-DVD discs available, for fear that they might get stuck with manufacturing capabilities for the wrong system. Suppose that during this period before the battle was resolved, a user decided to buy a portable Blu-ray disc player. At that time, the amount the user would have been willing to pay for the Blu-ray player would have been relatively low, because not much content was available, and if the war were to be resolved in favor of HD-DVD, not many discs would probably ever become available. So the user would have paid a relatively low price for the DVD player before much content was available. After the battle was resolved in favor of Blu-ray, more content started pouring into the market. Had the early adopter waited until after this new, greater level of content became available, he or she would have been willing to pay a higher price for the DVD player, reflecting the greater value of the disc player associated with the greater content availability. However, because the user bought the player before there was much content, the manufacturer of the disc player was not able to extract much value from the user. Instead, the user ended up retaining the surplus value associated with the subsequent increase in content availability. (Another possibility is that the content suppliers would have been able to extract some of this value through higher content prices.)

Had the user waited and paid a higher price, the manufacturer would have been able to extract a larger portion of the eventual value associated with the system. But since the user bought the hardware before the value was created, the manufacturer was unable to extract much of the eventual value. This is not to say that the price of the hardware will start out low and increase as more content becomes available. As we saw in the discussion of production lifecycles in Chapter 2, hardware prices will generally start out high upon introduction, then decrease as economies of scale and the effects of learning by doing are achieved, and later the effects of competition are felt. However, with more content available early on and the greater associated value it brings users, HW can start out charging higher prices, or maintain higher prices for a longer period of time, than they can when less content is available.

With more content available early on and the greater associated value it brings users, HW can start out charging higher prices, or maintain higher prices for a longer period of time, than they can when less content is available.

More generally, to be able to actually get at the profits associated with the system, HW need sufficient amounts of content to be generated *early on*. If earlier adopters purchase the hardware before the content becomes available, then HW will generally be able to extract only a relatively small amount of the value that content may eventually end up providing to users. This phenomenon can be seen in Figures 5-10A and B. It is only after much content becomes available that HW can start to generate significant profits on current and future sales of hardware.[10]

5.9 General Implications

I have just examined what hardware manufacturers' and software providers' profits look like under different network configurations (LL, LH, HL, and HH) and market scenarios. I have also explored how their profits change under alternative speeds of

10. Still, if content does not become available early on and HW miss out on generating much profit from sales of early hardware, HW might still be able to capture more value from later generations of the hardware.

new system adoption. Finally, I have scrutinized the sensitivity of their profits to changes in the various market drivers. Now that I have assessed these particulars, let me take a step back and look at the implications of the structure of the Hardware–Software Game more generally on technology system profits for both parties.

5.9.1 Overall Profit Potential

The profit potential of the new technology system is generally much greater and much more variable for HW than it is for SW. HW live or die with the success or failure of their new systems. If their new systems are adopted, HW stand to potentially earn very large profits. Conversely, if the new systems fail to take hold, then HW could lose large sums from unrecovered investments. SW, on the other hand, may very well earn greater profits with new systems than with older ones. However, the difference in profits SW will earn if the new systems succeed versus if they fail is not nearly as large as that for HW. This is because SW will benefit if the new system succeeds, but, unlike HW, SW always have the option of serving the market for the older technology systems if the new systems fail.

We can see the differences visually in the set of graphs in Figures 5-7A through 5-7D, by comparing the two-period combined profits for HW to those of SW for the areas in which users and content stay with the old technology as opposed to where users and content adopt the new technology. For HW, success (both users and content move to the new technology system) means very large profits, whereas failure (both users and content stay with the old technology system) means large losses. For SW, on the other hand, success means only moderately greater profits than those generated with failure.

5.9.2 Profits and Market Environment

HW's potential profits are very sensitive to the market and technology environment: the market potential of the new system (the innovativeness of the technology relative to available alternatives), the sensitivity of users to price (which is dependent upon the alternatives available for the new system), the strength of network effects associated with the new system, and the degree to which established users continue to buy new software (content saturation). A more favorable market environment provides a significantly greater profit opportunity for HW than a less favorable market environment. For HW, a more favorable environment means an innovative new technology, few alternatives, low user sensitivity to price, high direct network effects, moderate or high indirect network effects, and low tendency for content saturation.

For SW, a more favorable market environment is similar to that for HW, with the possible exception of a desire for high, rather than moderate, indirect network effects. Yet, the potential profits of SW do not vary that much with the new system. If the market environment for the new system turns out to be very favorable for SW, then they can ride along with HW. If, however, the new system does not provide a good profit opportunity (that is, the new system fails to take hold), then SW can simply fall back onto the older systems to generate profits. Figures 5-7A through 5-7D, 5-11, and 5-12 do, however, illustrate quite clearly that although new technology systems offer the potential for greater profits, both cost effects (it is less expensive to serve one market of users than two) and network effects lead to greater profits for SW when all users are in one place (at either of the two peaks in the diagram), even if that place is with the old technology system, rather than having some users with the old system and some with the new (between the two peaks).

Similarly, the amount of profits HW will earn is dependent upon the path of the game, that is, the number of users who adopt the new system each period and the amount of content that is available each period for the new technology system. If too few users adopt and/or too little content is available early on, then the momentum necessary to carry the new system over to adoption may never be created. On the other hand, too much momentum too early on may hamper the profit potential of HW, by preventing them from being able to extract the full value their systems create. Clearly, though, too much momentum too early on is not nearly as problematic as too little momentum. For SW, however, again, because they can moderate between the old and the new systems, their profits do not vary nearly as much with the path of the game.

5.9.3 Timing of Profits

The dynamics of the game suggest that HW will tend to generate losses during the early and late stages of adoption, but they stand to earn tremendous profits during the middle stages of the game, after early adopters have adopted the new system and the majority starts to follow suit. SW, on the other hand, face much less variability in profits from period to period, since they have the option of hedging by supplying users of the old system. It follows that HW will tend to concentrate their efforts on doing what will optimize profits during the middle stages of the game, whereas SW will simply concentrate on current-period profits.

5.9.4 Network Effects Dynamics

If the value of the system lies in the hardware itself, and network effects are weak, then the hardware part of the technology system must provide sufficient value to users before they will adopt. Conversely, if the value associated with a system lies in its network effects, so that the value of the system lies in its connectivity of users and/or its availability of content, then there have to be enough users or content for the new system before it will become valuable enough to users to make adoption worthwhile. Since the technology systems we care about here tend to have strong direct or indirect network effects, it follows that HW, and to some extent SW, will focus on network creation and network value more than on promoting the value of the technology per se.

Both HW and SW benefit from technology systems whose value lies in their networks of users; that is, direct network effects are strong. With strong direct network effects, users themselves create value; the more users there are, the more attractive the system is for new adopters (which benefits HW) and the greater is the pool of customers for SW.

When technology systems need a lot of software to generate value for users, HW may benefit from the extra value created for the system by the availability of content. At the same time, however, if the largest part of the value of a system for users is content, then HW could end up being worse off, because they will be at the mercy of SW for providing content to make the system valuable to users. As for SW, when the technology system needs a lot of content to create value for users, then SW are better off. In this case, it is their content that creates value for the system, and so user demand for content will be larger, for which users will pay.

The last aspect of network effects is the potential for content saturation. When new content complements (rather than substitutes for) existing content—that is, new content provides new and different applications for users—SW face a more profitable environment than one in which new content is really just more of the same as existing content. HW, on the other hand, prefer environments in which there is a moderate potential for content saturation. In this case, users of older technology systems are more likely to upgrade to new systems sooner rather than later. Furthermore, when new content competes more directly with existing content, content prices fall, making the system less expensive to users, which encourages new adoption by users who have yet to adopt.

5.9.5 Value Extraction

Recall that I posed a question in Chapter 1: When a network grows, say, from 100 members to 1,000 members, who gets to keep the extra value to the original 100 members created by the addition of the 900 new members? After having performed all the analyses described in this chapter, I can now answer that question. If HW and/or SW can extract the value through higher prices to original members, then the providers will get to keep that value. If, however, providers cannot raise their prices to the original members as value increases, then the users will get to keep that value.

Under leasing or subscription programs, providers might be able to raise the price to existing members as network size grows. However, if hardware and software components are sold rather than leased, providers cannot generally go back to existing users and demand that they pay surcharges if and when the value of the system has increased (thank goodness!). In this case, providers could try to extract value from early adopters by charging them prices based on expected future network size. But this does not usually work. This type of pricing will tend to cause users to delay adoption, yet this is precisely the time when network providers are trying to encourage adoption. In the end, when components are sold rather than leased, it is usually existing users who get to keep the value associated with future growth in network size.

Well, then, if providers cannot extract value from the original 100 members of the network, how can they extract the most value from the 900 new members? We have established that users will generally be willing to pay only lower prices associated with current network sizes, and not higher prices associated with larger expected future network sizes. As such, providers will extract the most value from new adopters by repricing their network components as frequently as possible; that is, rather than signing up all 900 new members at once for prices based on the current network size of 100, providers want to charge the 101st member of the network a price reflecting an existing network size of 100, the 102nd member a price reflecting an existing size of 101, the 103rd member a price reflecting an existing size of 102, and so on.

Of course, the fact that there are other transaction costs associated with changing prices means that providers will not change their prices for each new member. However, the idea is that neither HW nor SW want too much hardware or software to be sold too quickly. If system adoption happens too quickly, then providers will not have a chance to reprice their hardware or content so as to fully reflect their value to users, and it will be the users themselves who will retain most of the value associated with network effects.[11] In this case, the market will become saturated before providers are able to extract the full value of the network effects being generated.[12] When HW can change the price of hardware as the size of the network grows, they will be able to extract a greater portion of the value of the direct (and indirect) network effects.

Figures 5-10A and B exemplify these effects. HW want enough users to adopt their technology systems each period so as to sufficiently increase the value of adoption for the next period's adopters. In Figure 5-13, this is seen by the fact that HW's optimal rates of adoption tend to be 10 to 30 percent, rather than 1 percent. At the same time, however, HW do not want too many users to adopt their technology systems each period, for this will prevent them from being able to extract the full value of increases in the size of the network; that is, optimal rates of adoption tend to be less than 40 percent.

Finally, the last generalization to be made about value extraction is that HW want SW to supply as much content as possible at the lowest prices possible to create value for the system. This will enable HW to charge higher prices for hardware and extract value created by SW. Likewise, SW will earn greater profits when HW charge lower prices and SW can extract the excess value created through higher content prices.

---

11. On one hand, hardware manufacturers would like to be able to continuously reprice their hardware to reflect the constantly increasing value of the system as new users adopt. On the other hand, if users anticipate an imminent decrease in price, many will choose to delay adoption until the price is more favorable. (Economides and Himmelberg [1995] present a perfect example of users waiting to adopt until the price stopped dropping in the case of the adoption of fax machines in the United States.) In this case, then, longer periods between price changes will increase the opportunity costs to users of waiting, thereby encouraging them to adopt sooner. The ideal period between repricing will balance these value-added and delay-increasing effects.

12. Judson Coplan (2006) suggests that "the remarkably rapid penetration of DVD players across the U.S." caused the market to become saturated too quickly. Coplan noted that "there are numerous reasons for the rapid penetration of DVD hardware, such as inclusion of DVD players in PCs and gaming consoles, an easy-to-grasp technology transition from VCRs, and rapidly falling prices for the players themselves." He then cites Scott Hettrick, editor-in-chief of trade publications Video Business and DVD Exclusive: "The industry shot itself in the foot by lowering DVD prices too much and too quickly."

CHAPTER SIX

Addressing the Chicken-and-Egg Problem

I now have a clear understanding of the structure and dynamics of the Hardware–Software Game. I understand who the players are, what incentives they face, and which actions they are led to take. With this understanding, I can now address the issue of how one set of players, hardware developers in particular, can change the incentives of the other sets of players, that is, users and content providers. By taking action to change users' and content providers' incentives, hardware developers can induce these two sets of players to change their actions to better suit the needs of hardware developers, namely, to switch over from the old technology system to the new system as quickly as possible.

Ultimately, inducing mass adoption of a new technology system becomes the burden of hardware manufacturers, because it is the hardware manufacturers that have the most to gain. Mass adoption of new technology systems will enable hardware manufacturers to generate profits associated with sales of new technology hardware to users. In contrast, successful adoption of new systems will enable content providers to earn only incremental profits, that is, profits associated with the provision of content for the new technology systems *above and beyond* profits they could have earned by continuing to supply content for the existing technology systems. If hardware manufacturers' new systems fail to succeed in the marketplace, content providers can still continue to supply content to users of existing systems. Similarly, users who choose to upgrade to new systems will realize the incremental gains the new systems have to offer over older systems, which are generally small relative to the profits hardware manufacturers stand to generate. It follows that to the extent that adoption of new technology systems faces resistance, it will be up to hardware manufacturers to make the effort to encourage users and content providers to accept new systems.

An important point to note is that it will always be better for hardware manufacturers to understand how the chicken-and-egg problem comes into play *before* they develop their new technology hardware, rather than waiting until after they develop their new hardware and are ready to introduce it into the marketplace. By understanding the issues in advance, hardware developers give themselves many more options to set the stage to increase the likelihood of successful adoption of their new technology systems. Waiting until after their hardware has been developed and is ready to hit the market, on the other hand, will severely limit their options, by forcing them to be more *reactive* to the market, given the nature of their already developed technology systems, rather than having the luxury of being *proactive*, to adapt their new technology systems to the needs of the market.

In this chapter I first reiterate the problem I am addressing, that is, the chicken-and-egg problem associated with the introduction of new technology hardware, which impedes adoption of new technology systems. I next discuss general actions hardware developers can take to overcome the chicken-and-egg problem or otherwise increase the speed of adoption of their new technology systems by users and content providers. After that I discuss scenario-specific actions hardware manufacturers can take to increase the speed of adoption of new technology systems by users and content providers.

6.1 Statement of the Problem

The most vital segment of new technology adoption occurs from its introduction into the market through the point at which either technology adoption reaches critical mass or the momentum dies and the technology exits the market. Accordingly, I will focus my examination of the underlying dynamics of the game on the early stages of adoption, when early adopters and the early majority are considering making the switch from older technologies to next-generation systems.

When a new technology is introduced into the market, as long as the technology system itself is compelling and shows promise, many innovators and early adopters will adopt, despite the current lack of user or content networks. Not to trivialize the

The most vital segment of new technology adoption occurs from its introduction into the market through the point at which either technology adoption reaches critical mass or the momentum dies.

difficulty of introducing new technology systems into the market, but getting early adopters to buy into the new technology per se is not generally the hardest part. The hardest part is getting past the early adopters and making the value proposition associated with the new systems compelling enough to make adoption by the early majority worthwhile. Creating a convincing case for adoption by the early majority requires that hardware manufacturers establish *both* the desirability of the new technology systems themselves and their staying power; that is, hardware manufacturers must demonstrate that the new systems are not only better than the older systems, but that they are also here to stay and will form the basis of, or otherwise play a significant role in, future ecosystems. In other words, the early majority requires solid validation of new technology systems before it will commit to upgrading. Solid validation takes the form of large actual or expected installed bases of users and/or large actual or expected availability of content.

Creating a convincing case for adoption by the early majority requires that hardware manufacturers establish *both* the desirability of the new technology systems and their staying power.

Noteworthy is the fact that this validation does not require that the user/content networks actually be in place, only that they are expected to be in place very soon. However, the lower the costs of waiting to adopt are, the more likely the early majority is to wait until sufficient networks have already been established before it commits to upgrading. These dynamics mean that there will generally be a time lag between the adoption of the new technology by early adopters and by the early majority as the early majority waits for validation.[1]

So, the early majority needs a sufficiently established technology/network package to have been established before it will adopt the new system. But why can't hardware manufacturers set very low prices on initial sales of hardware, or just give their hardware away for free, until critical mass has been achieved, then make up any lost profits on subsequent sales to post-critical-mass adopters?

1. In *Crossing the Chasm,* Geoffrey Moore (1999) essentially makes this point, where the chasm is the interval between acceptance of the new technology by early adopters and adoption by the early majority.

This type of strategy most commonly works only when both the production and sale of hardware and software/content are controlled by the same player. In other words, this is the classic Gillette model, in which joint profits on sales of razors and blades are maximized by selling the razor at a very low price and making up lost profits on sales of the blades. Younger technologists will probably recognize this model better as the printer and ink model, in which makers of computer printers sell printers at very low prices but then generate large profits on subsequent sales of ink. If, however, the supplier of the hardware and the supplier of the software for a technology system are two independent entities, each of which must maximize profits from sales of its own component of the technology system, then the Gillette model does not work. In this case, the Gillette model will usually result in hardware manufacturers losing money on sales of their hardware components, while content providers collect the entire value of hardware–software systems through sales of content. One could propose that in this case content providers could share their profits with hardware manufacturers, but then this would no longer be the scenario in which hardware manufacturers and content providers are independent entities, each acting in its own self-interest.

More generally, the strategy of giving away just enough hardware to enable hardware manufacturers to achieve critical mass, and then making up any lost profits on subsequent sales of hardware to post-critical-mass users, would hold potential only when three conditions are met simultaneously: (1) the portion of total eventual users who must adopt the hardware in order for hardware manufacturers to achieve critical mass is relatively small, (2) critical mass can be achieved relatively quickly, and (3) post-critical-mass users are willing to pay a relatively high price for the new technology hardware.

The first criterion required for a give-the-product-away-until-critical-mass-is-achieved strategy to be profitable is that the portion of total users required to adopt

> A strategy of giving away hardware until critical mass is achieved would hold potential only when three conditions are met simultaneously:
>
> - The portion of users who must adopt the hardware to achieve critical mass is small.
> - Critical mass can be achieved quickly.
> - Post-critical-mass users are willing to pay a high price for hardware.

the new technology before it reaches critical mass is small. Visually, this means that in Figure 6-1, the lightly shaded region to the left is small compared to the more darkly shaded region to the right.

This follows from the fact that as more people are required to adopt the technology before critical mass is achieved, fewer users will be available from which to generate sufficient revenues to make a profit. The number of users required to achieve critical mass tends to vary across technologies; however, the number will tend to be higher when the costs of technology adoption are higher, which happens when (1) the price of the new technology is higher, (2) the switching costs are higher, and/or (3) the risks of being stranded are higher. So, the number of adopters required to reach critical mass will be relatively small when the costs to users of new adoption are relatively low.

The second criterion for a give-the-hardware-away strategy to work is that critical mass can be achieved relatively quickly. Time costs are high in markets in which the pace of technology evolution is quick. The rapid rate of introduction of new competing technologies means that any given technology has only a short window of time in which to penetrate the market and generate revenues for technology suppliers, before new and better technologies come along and steal away users. So the longer suppliers must wait to achieve profitability, the lower the likelihood is that they will eventually be able to do so before the market moves on to other technologies. In Figure 6-1, this criterion means that t* is relatively small compared with T.

The last criterion required for a give-the-product-away-until-critical-mass-is-achieved strategy to be profitable is that hardware manufacturers must be able to charge post-critical-mass adopters high enough prices for system hardware to make market entry a profitable proposition. The problem here is that the early adopters who are getting

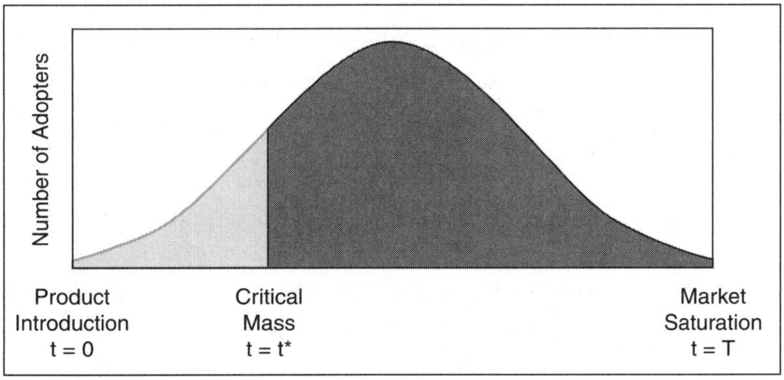

Figure 6-1 Technology adoption lifecycle

the technology for free under this strategy are generally those users who would be willing to pay the highest price for the new technology. At the same time, the post-critical-mass users from whom large revenues must be generated are precisely those users with the lowest willingness to pay. This situation generally precludes a give-the-product-away-until-critical-mass-is-achieved strategy from being profitable, unless the technology system eventually becomes adopted by a very large number of users, in which case only moderate prices need to be charged for hardware manufacturers to recoup their investments.

Given the low probability of having all three of these criteria met, it follows that when hardware manufacturers and content providers are independent players that act in their own self-interest to maximize profits, a strategy in which hardware manufacturers give away their products until critical mass is achieved will usually not be a profitable one.

So now we are left with a situation in which hardware manufacturers have introduced their new technology systems into the market, and they have achieved some penetration from sales to early adopters. However, there is not enough content yet available to make adoption of the new technology systems worthwhile for the early majority. At the same time, the installed base of new technology users is too small to make it profitable for content providers to redirect their resources into the production of new system content. We have thus slammed headfirst into the chicken-and-egg problem of new technology adoption.

The big questions are then: At the early stages of product introduction, when the small size of the installed base of new technology users discourages the provision of any substantial amount of new technology content,

- How do we encourage the early majority to adopt the new technology, despite the lack of content availability?
- How do we encourage the content providers to supply content for the new technology, despite the lack of a sufficiently large installed base?

6.2 General Responses to the Chicken-and-Egg Problem

I have established that the adoption challenge comes after early adopters have begun to adopt the new technology and before the early majority begins to do so. The big problem is twofold: getting the early majority to adopt as soon as possible, and getting content providers to switch over to the production of new system content as soon as possible.

The various issues facing the three sets of players, together with possible actions hardware manufacturers can take to overcome the chicken-and-egg problem, are summarized in Figure 6-2 and discussed in the following sections.

6.2.1 Engineer New Systems for Maximum Likelihood of Adoption

In Chapter 4 I used the case study of the experiences of Apple with its Newton and Palm with its Pilot to describe what makes a technology system attractive to users. The three key features are value, simplicity, and compactness—think "Zen of Palm."[2] Value means clear and material benefits over the alternatives. Simplicity means ease of understanding, ease of learning, ease of use, less time, less aggravation, less to go wrong. Compactness means lightness, portability, lack of bulk, lack of space, lack of intrusion. Together, value, simplicity, and compactness mean worthiness, elegance, low cost, low price, great appeal, quick adoption.

I have heard it said that having a great team is more important than having a great technology. However, having the foresight and making the effort to transform an innovation into a great technology will make the journey a whole lot easier.

6.2.2 Create Expectations of New System Adoption

Suppose you wanted to create the perception that there will be a shortage of, say, bottled water. You announce that a major bottling plant has had some sort of problem that will force it to close down. What will happen? In anticipation of the announced shortage, many people will run out and stock up on bottled water before the stores run out. However, as people buy up the stores' stocks of bottled water faster than the stores can restock, the people actually create the shortage they expected. *Voilà!* Self-fulfilling expectations.

As I have repeated throughout this book, both technology users and content providers have the option of staying with old technology systems, which provide known (that is, relatively risk-free) benefits to the respective parties, whereas moving to new technology systems requires both groups to take risks. Adoption of new technology systems requires them to make sacrifices in the short run: Users sacrifice money to buy new technology systems, and content providers sacrifice profits on sales of content for old technology systems. In exchange for these sacrifices, both parties *hope* to be more than compensated in the long run—users by achieving greater benefits, and

2. See section 4.1.

	Users [A]	Hardware Manufacturers [B]	Content Providers [C]
[1] Problem	Choose whether or not to upgrade to the new technology each period	Choose how much to charge for the new technology system hardware each period	Choose how much old vs. new technology system content to produce each period
[2] Value/Profits	• Early adopters want good technology • Early majority wants network (users/content) as validation of desirability and staying power • Want low risk/switching costs	• Margin vs. volume • Profits increase with size of installed base (value and cost) • Profits increase with content (value)	• New sales vs. cannibalism (market saturation) • Profit per user vs. number of users (cater to large installed base) • User access to content
[3] Actions	• Expectations about what others will do • Stay with old technology system until (value of upgrading) > (price + switching costs)	• Too little content early on => lose volume/margin • Want moderate adoption during early periods to create validation/installed base	• New technology content increases slowly with number of user upgraders
[4] Network Externalities	• Direct (network of users): value increases with actual/expected size of installed base • Indirect (network of users): early adopters validate technology for later adopters • Indirect (availability of content): value increases with actual/expected content availability	• Larger installed base increases value and enables higher price • Greater content availability increases value and enables higher price (subject to content saturation)	• Larger installed base increases value and enables higher price • Greater content availability increases value and enables higher price (subject to content saturation)
[5] Switching Costs	• Price of new system, installation costs, training costs • System downtime during transition period • Risk of being stranded with the new technology • Critical mass creates momentum (costs of not switching become high)		• Investment costs (new equipment, license fees, etc.) • Training costs • System downtime during transition • Loss of economies of scale
[6] Decrease Value of Old System	• Stop providing support for old system • Make new offerings dependent upon use of new system • Lobby for regulations/standards that make old systems obsolete		• Stop providing support for old system • Make new offerings dependent upon use of new system • Lobby for regulations/standards that make old systems obsolete
[7] Increase Value of New System	• Increase value of new technology ◦ Increase size of actual/expected installed base ◦ Increase amount of actual/expected new content • Decrease price		• Increase relative potential size of installed base for new technology system • Minimize cannibalism of new technology system sales • Increase profit per user for new technology system
[8] Decrease Switching Costs	• Decrease risk of nonperformance • Decrease downtime • Decrease installation/training costs		• Decrease investment/training costs • Decrease downtime

Figure 6-2 Hardware–Software Game dynamics

content providers by generating greater profits—which means the new system must become successfully adopted.

Obviously, then, hardware manufacturers should try to minimize risk in order to get users and content providers to adopt their new systems. But this appears to be a catch-22 problem: For hardware manufacturers to convince users and content providers to adopt new technology systems, they must first achieve critical mass. Yet they cannot achieve critical mass unless users and content providers adopt the new technology systems.

Actually, though, the trick is really one of creating *expectations* about adoption, rather than creating adoption itself. This follows from the fact that people act based not necessarily on what is actually happening, but rather on their perceptions or expectations about what they think is true—"truth is perception"—as in the bottled water example. Similarly, creating the perception that a technology system is being adopted or will imminently be adopted leads people to act as if the system is actually being adopted. So they jump on the bandwagon and adopt the system, which then leads the system to become adopted. In other words, the trick is to create self-fulfilling expectations of technology adoption.

More specifically, if users believe that other users are adopting or will soon adopt the new technology system and that content will soon become available, then the expected value to them of adoption will be that associated with having a large base of users and a lot of content available. Likewise, if content providers believe that users are adopting or will soon adopt the new system, then they will think that demand for new system content will be high and that demand for old system content will be low. With these expectations, content providers will do best by switching resources over to the production of new system content to cater to the growing demand.

There are several actions hardware manufacturers can take to create these expectations of imminent technology system adoption. Establishing new systems as standards is perhaps one of the most straightforward (though not necessarily easy) means hardware manufacturers can use to establish that their new systems will be adopted by the market. "In standards wars, there is a very real sense in which the product that people expect to win, will win. Nobody wants to be stranded with an incompatible product, so convincing potential adopters you have the winning standard is critical."[3]

A second useful technique for convincing the market that system adoption is inevitable is for hardware manufacturers to use "expectations management," such as

3. Varian 2001, 87.

advertising, product preannouncements, or market share claims, to persuade users that their systems will become dominant. We can learn a lot from teenagers, who are experts at managing their peers' expectations: "But, Mom, everyone else is going to the party!"

A third approach is for hardware manufacturers to take actions that "show" (that is, signal) the market that they are committed to the new technology systems, generally by making large investments in or commitments to continued supply of the new systems. For example, manufacturers can build out capacity to produce the new technology hardware. The telecoms did this when they built out their extensive networks of fiber-optic cables. When the telecoms built their networks of cables, we knew that there would be plenty of network capacity well into the future. Alternatively, manufacturers can create large distribution networks,as Amazon, Netflix, and FedEx, among others, have done. Furthermore, when we see a large build-out of fuel stations to accommodate vehicles powered by alternative fuels, such as ethanol or hydrogen, we will know that the technology is here to stay. In other words, hardware manufacturers may be able to speed up the pace of adoption of their new technology systems by committing to be in the marketplace well into the future. Such a commitment may reassure new adopters that new technologies will continue to be used and supported well after they have adopted them. Commitments to having a future presence may be made, for example, by building out capacity or infrastructure or making other large up-front investments in future production to guarantee the new technology system will be around indefinitely.

A fourth approach is for hardware manufacturers to convince large market players, either large users or large providers of complementary products and services, to publicly commit to the new technology, that is, *get buy-in* from large users or suppliers. This technique was recently used in the DVD format wars. Toshiba secured alliances for its HD-DVD format with Microsoft and Universal, while Blu-ray developer Sony secured alliances with Sony, Panasonic, Walt Disney, Twentieth Century Fox, and Lionsgate Entertainment.[4]

To summarize, hardware manufacturers can speed up the rate of adoption of their new systems by creating the perception that their systems are being adopted or will imminently be adopted in the market. Innovators can create expectations of adoption by (1) establishing new systems as standards, (2) using expectations man-

4. Wingfield 2006b.

agement, (3) signaling to the market that they are committed to long-term supply, or (3) getting buy-in from large users or suppliers.

6.2.3 Establish a Good Reputation

General Electric (GE) has a good reputation in the business world. It has been around since the late 1800s, and since that time it has successfully introduced new and innovative technologies again and again. Examples of GE innovations include the carbon filament, incandescent lighting, the central power station, the X-ray machine, the electric fan, the vacuum tube, refrigeration, the transoceanic radio system, and the washing machine.[5] Providers of accessory products and services face relatively minimal risk that new technology systems introduced by GE will not become successfully adopted in the marketplace and so should generally be eager to provide complementary products for GE technology systems.

Hardware manufacturers can attract both users and content providers early on to adopt their new technology systems by establishing good reputations for themselves as companies with high-quality products, good service support, and/or staying power. Establishing good reputations decreases the risks to users and content providers that the companies' new technology systems will fail to take hold or otherwise leave early adopters stranded; so, they will be more eager to adopt new technology systems earlier, rather than waiting until the systems gain more traction.

6.2.4 Decrease Relative Value of Old Systems

As I illustrated in Chapter 3 with the numerical example of user adoption of new system hardware, users generally have existing systems in place that provide benefits similar to those offered by hardware manufacturers' new systems. Users will adopt new systems only if the value they provide *above and beyond that of existing systems* is worth the total cost of switching (new system price plus switching costs). Likewise, content providers will switch their resources over to the new system only when doing so is more profitable than staying with existing systems. So system innovators can effectively increase the net value of their new systems both to users and to content providers by decreasing the value of existing systems.

5. www.ge.com/innovation/timeline/index.html.

6.2.4.1 For Users

When high-definition television sets first started hitting the market, I remember thinking to myself, "I have a really nice television set, with a nice large screen and good picture quality, that has served me perfectly well for the past five years. It wasn't cheap when I bought it, and the new HDTVs are really expensive. The extra sharpness associated with HDTV is just not important enough to me to get me to throw away my perfectly good TV and replace it with an HDTV."

More generally, users will be much more reluctant to upgrade to a new technology if the old technology still provides a lot of value. The *extra* value the new technology provides *above and beyond* the value of the old technology must be worth the entire price of upgrading, which just doesn't happen very often. And the greater or more expensive the value of the old technology system is, the more difficult it will be for the new technology to provide incremental benefits that exceed its price. At the same time, the greater or more expensive the value of the old system is, the riskier the upgrade will be. Both of these issues tend to cause more expensive technology systems to be more slowly adopted than less expensive ones.

There are a few ways hardware manufacturers can reduce the value of old technology systems to users, thereby encouraging them to upgrade to new technology systems sooner than they might otherwise. First, hardware manufacturers can stop providing support for the old technology hardware, for example, by discontinuing the provision of patches, upgrades, parts, or services. "Discontinued GM Camaro/Firebird Parts: As you may already know, car manufacturers do not manufacture replacement parts indefinitely. Eventually, they cease to make any more components for older vehicles."[6] Second, hardware manufacturers can make all new product and service offerings dependent upon the use of the new technology system. Examples of this strategy at work include later generations of Word or Excel or other applications that require more memory to use than earlier PCs were equipped with, CD players that do not have the ability to play records or cassette tapes, or DVD players that do not have the ability to play VHS tapes. Another example is when Intuit released a new version of its Quicken software and forced users to upgrade to the new version by preventing them from accessing new financial statements with the old software.[7] Third, hardware manufac-

6. http://tpiparts.net/discontinued_parts.

7. www.discourse.net/archives/2004/05/annals_of_software_obsolescence_intuit_is_even_more_evil_than_i_though.html.

turers can lobby for regulations that make the old technology systems obsolete. For example, hardware manufacturers can lobby to have minimum quality or safety standards raised to levels above those met by the old technology systems, such as mandating that they be lead-free, earthquake- or hurricane-proof, environment-friendly, handicap-accessible, or equal-opportunity (multilingual or multicultural).

6.2.4.2 For Content Providers

The same strategies hardware manufacturers can use to reduce the value of old systems to users will also end up reducing the value to content providers of continuing to supply the market for the old systems. If hardware manufacturers stop providing support for old systems, make new offerings dependent upon the use of newer systems, and/or have standards or regulations established that are not met by older systems, then users will be reluctant to continue to purchase content and accessories for older systems, knowing that they have become obsolete. This means demand for old system content and accessories will dry up, leaving content providers without a market to cater to.

6.2.5 Increase Relative Value of New Systems

Clearly, users and content providers will be more likely to switch over to newer systems when their value is much greater than that of older systems. System innovators can thus speed the rate of adoption of newer technology systems by making the new systems as valuable as possible to users and content providers.

6.2.5.1 For Users

For users, the relative value of the new technology system will be higher, and so users will be more likely to adopt it, when (1) the value of the technology itself, relative to that of the old technology, is higher; (2) the value associated with direct network effects is higher; that is, the size of the current and expected future installed base for the new technology, relative to that of the old technology, is higher; and/or (3) the value associated with indirect network effects is higher; that is, the current and expected future availability of content for the new technology, relative to that for the old technology, is higher.

Value of the Technology Itself Refer back again to the example of my reluctance to adopt HDTV. As technology systems evolve, it becomes more and more difficult for manufacturers to come up with improvements on older systems that make users want to upgrade immediately. More often than not, when a next-generation system comes out, older systems being used are still working, if not perfectly, then at least at a level that provides users with enough value to make an immediate upgrade seem wasteful. In these cases, users will often wait until the benefits of their older systems degrade for example, when their service contracts expire or when old system components wear out or die—before upgrading to next-generation systems.

Hardware manufacturers can increase the pace of new system adoption by making the added benefits of their new systems as valuable as possible relative to those of the old technology, as was discussed in section 6.2.1. The goal is to make the *added benefits* of new systems so valuable that users will rush out and upgrade to the new systems before waiting for their older systems to lose value. For example, when cell phones came onto the market, they offered a whole new dimension of value to the telephone: the ability to be away from home or office and still be able to make and receive calls. This added feature made it justifiable for users to immediately go out and purchase portable telephones for use while away from home, even if they had just bought new cordless phones for home or office use. As another example, faster technology and greater availability of content led many users of the original iPhone to upgrade to 3G as soon as the newer version was released only 18 months later:

> *The iPhone 3G, which sold more than 3 million units after just a month on the market, is giving techies plenty of reason to part with their 2G and 1G models. With faster wireless technology, GPS mapping, support for enterprise features like Microsoft Exchange, and the new App Store, the new iPhone 3G has taken control of the market and inspired clones from the world's other leading high-tech companies … "Other companies have been racing to catch up with Apple, which shows the popularity of the product and why so many people are willing to upgrade to a 3G so soon after purchasing a 1G or 2G model."*[8]

One of the challenges associated with technology evolution is that the more valuable hardware manufacturers make their next-generation technologies, especially when they improve product durability, so as to maximize the likelihood of rapid product adoption, the more difficult it will be for them to come up with ever better product features that will make it worthwhile for users to quickly upgrade to future generations of their products. But, of course, it is this competitive challenge that drives technology evolution.

8. *PRWeb* 2008.

Size of Installed Base After making the value of the technology itself as great as possible in order to speed up the pace of adoption of new systems, the second possibility is to make the value of direct network effects—that is, the size of the actual and/or expected installed bases—larger than that of older systems. Establishing a new system as a standard (actual or de facto) is a pretty clear-cut means of ensuring a large installed base for a new technology. Qualcomm established CDMA as a mobile phone standard, which ensured Qualcomm a large share of the chip market for mobile phones. Sony recently established Blu-ray as the de facto standard for high-definition DVD, which will ensure it a large share of the next-generation market for DVD players.

Research in Motion's (RIM) BlackBerry was one of the first devices to offer mobile e-mail services (that is, cellular access to e-mail). RIM could have established its own network of users for its technology system, by enabling BlackBerry users to send and receive e-mail messages only to and from other BlackBerry users. However, this would have significantly limited the value of the system and made it much more difficult for RIM to build up a network of BlackBerry e-mail users. Instead, RIM made the e-mail services offered on its BlackBerry devices compatible with all other e-mail services, so that BlackBerry users could send and receive e-mail to and from any other e-mail users, be they using BlackBerrys, PCs, or smart phones. RIM's choice to make its e-mail services compatible gave it a smaller piece of a larger network (all e-mail services), as opposed to a larger piece (actually, the whole piece) of a smaller network (e-mail services to and from BlackBerry users). Yet, this decision also enabled RIM to build up its network of users and achieve critical mass, which it might not have been able to do had it insisted on creating a proprietary network. In fact, the greater value associated with being able to send and receive e-mail to and from any other e-mail users, as opposed to just other BlackBerry users, made the device more valuable to new adopters, thereby enabling RIM to charge a higher price for its hardware. So RIM's choice to make its system compatible with others not only enabled it to generate higher profits, but it also created a greater likelihood of achieving critical mass. Not a bad decision!

So we see that an alternative way of ensuring a large base of users for new systems is to make the systems compatible with existing networks. With backward and/or cross-product compatibility, user networks for new technologies will effectively include installed bases of all compatible technologies. Of course, innovators of new systems that are compatible with existing systems are not able to generate revenues from users of the existing (other) technology systems. However, by making their new systems compatible with existing systems, system innovators can greatly alleviate the problem of achieving critical mass on their own. What's more, by making their new systems compatible with existing systems, system innovators can also benefit from the higher network values for their own systems immediately created by plugging into existing networks.

Establishing compatibility with other networks may or may not require hardware manufacturers to pay fees to access other networks. Also, cross-network compatibility will have greater appeal to newer or smaller-sized manufacturers, which have more to gain from being able to link into other (larger) networks than larger or more established manufacturers. Yet smaller hardware manufacturers may have less leverage, and thus be less likely to be able to achieve buy-in to larger-sized networks, than larger manufacturers, who may need the buy-in less.

A third means of increasing the probability of establishing a large base of new system users is to make the system multifunctional, so that it draws from a wider base of users than older technologies. For example, by combining its mobile e-mail services with mobile telephone services, Internet access, and PDA features, RIM attracts users from four distinct markets: PDAs, cell phones, e-mail services, and the Internet. As another example, by combining copy machines, facsimile machines, and printers into all-in-one units, hardware manufacturers simultaneously draw from three distinct user markets. As a third example, Sony (PlayStation 3) and Microsoft (Xbox 360) designed their gaming systems to be compatible with HDTVs.[9] Manufacturers of HDTVs are thus able to draw from markets of television viewers, but also from users of gaming systems.

One more means of ensuring a large base of new system users is to use expectations management, such as advertising, product preannouncements, or market share claims, to persuade users that one's system will become dominant. Expectations management is more fully discussed in section 6.2.2.

To summarize, hardware manufacturers can speed up the pace of adoption of their new systems by making the actual and/or expected sizes of installed bases for new systems larger than those of older systems. This may be accomplished by (1) establishing the new systems as standards, (2) making new systems compatible with existing networks, (3) making new systems multifunctional so that they draw from wider bases of users than older technologies, and (4) using expectations management.

9. Wingfield 2006a.

Availability of Content Consider the case of digitization of medical records. When all medical records in the United States eventually transition from paper to electronic files, the medical establishment will benefit from all kinds products and services relating to the creation, maintenance, and management of medical files, such as video management technology for handling X-rays, CAT scans, and the like; medication management technology for managing drug prescriptions and protecting against toxic interactions; of course, insurance, billing, and payment management technologies; security technology, preventing medical data from falling into the wrong hands; document management technology for organizing correspondence among physicians; and so on. While suppliers of these medical management technologies are developing, they would all mutually benefit from forming alliances in support of the general trend toward digitization.

This example illustrates how system innovators can help increase the appeal of their new systems, and thus the pace of adoption, by "building alliances with suppliers of complementary products."[10] In other words, to the extent that complementary goods and services are not included in the offerings embedded in our definition of "content provider," hardware manufacturers can construct ecosystems around their new technology systems by partnering with suppliers of related products and services.

Apple is king in creating ecosystems around its technology systems. When the iPod was being developed, Apple recognized that the users would not be willing to pay much for the iPod, or adopt it willingly, unless or until there was an easy source of legal digital music with which users could stock their iPods. Since no source existed at the time, Apple engineered the development of iTunes. iTunes was introduced on January 9, 2001, at the MacWorld Expo: "Apple® today introduced iTunes, the world's best and easiest to use 'jukebox' software that lets users create and manage their own music library on their Mac®."[11] The first-generation iPod was released later that year, on October 23, 2001. With an immediately available, cheap, and easy means of stocking the iPods with music, the iPod was instantly a huge success. Apple applied the same concept to its 3G iPhones by creating a developers' program to encourage the supply of applications for the iPhone by third-party suppliers:

> *Apple is getting ready for one of its biggest events this year: the launch of the 3G iPhone and its already more than famous AppStore ... the AppStore is a concept that could*

10. Varian 2001, 87.

11. Apple Computer, Inc. 2001.

> *make the difference. Steve Jobs … announced the AppStore back in March and it is basically [a] new application that lets users browse, search, purchase and wirelessly download third party applications directly onto their iPhone or iPod Touch.*[12]

By facilitating this ecosystem of applications, Apple effectively ensured that users would eagerly adopt its 3G system upon release.

More generally, methods that hardware manufacturers can use to help persuade content providers to switch over to the production of new technology content, thereby speeding up the pace of adoption of their new systems, is discussed in the following sections.

Price of the New Technology System Certainly, users are more likely to adopt systems that are less expensive. One of the best ways system innovators can ensure they will be able to offer lower prices for technology components is by engineering their systems as efficiently as possible. This strategy was used effectively by Palm in designing its Pilot, as we saw in Chapter 3. Another example of a company that successfully engineered its operations to be as efficient and low-cost as possible is Southwest Airlines. In particular, Southwest purposely avoids flights to and from larger airports, which have higher associated costs of use than secondary airports. Southwest also minimizes the turnaround times for its planes (the time its airplanes stay on the ground between flights). It does this, for example, by not having reserved seating, which enables it to load passengers more quickly onto its planes. With shorter turnaround times, Southwest's planes spend less time idling on the ground and more time in the air generating revenues. As yet another example of its purposeful, efficient system engineering, Southwest formed its fleet of airplanes all of the same model so as to economize on maintenance fees—fewer parts to stock, less training for mechanics, and so on.

As an alternative to engineering their systems to have the lowest price possible, hardware manufacturers may decrease their prices to early adopters, thereby speeding up the pace of adoption, by using *penetration pricing*. Penetration pricing entails setting an unprofitably low price early on, so as to amass an initial base of users, before raising the price to a more profitable level later on.[13] I discussed earlier how a strategy in which hardware manufacturers give away their products for free until critical mass is achieved is usually not profitable, and the same applies to product discounting. Yet there are

12. Brenn 2008.

13. Wikipedia provides a good description of penetration pricing: http://en.wikipedia.org/wiki/Penetration_pricing.

some circumstances in which penetration pricing may be profitable, such as if adoption by a few key early adopters is enough to kick-start the viral phenomenon of adoption.

6.2.5.2 For Content Providers

For content providers, the relative value of the new technology system will be higher, and so they will be more likely to switch their resources over to the production of content for the new system, when (1) there are more users of the new system—that is, the size of the installed base is larger; (2) sales of content to new system users does not come at the expense of sales to old system users—that is, content providers could not have sold the same content to old system users that they instead sell to new system users; and/or (3) the per-unit profitability of new system sales of content is greater than that of old system sales.

Size of the Installed Base By providing incompatible gaming technologies, Microsoft (the Xbox), Sony (the PlayStation), and Nintendo (the GameCube and Wii) have partitioned the video game market into three distinct submarkets. As a result, rather than being able to supply video games to a single market of all users, suppliers of video games and gaming accessories must choose one particular submarket in which to participate; that is, with three incompatible submarkets, gaming market suppliers face only about one-third of the potential profits they would have stood to earn had the market comprised a single standard. These smaller markets most likely lead suppliers to provide fewer games and accessories overall than there would be in a single, unified market.

Hardware manufacturers can encourage content providers to supply content for new technology systems by increasing the sizes of the markets for new systems relative to those of the old systems. One way to accomplish this is for hardware manufacturers to create single, unified markets for new technology systems, either by establishing standards or by making new technology systems compatible with existing systems or standards. In the previous example relating to the market for video games, if a next-generation manufacturer could provide a technology that unifies the separate systems offered by Microsoft, Sony, and Nintendo, it could quickly attract suppliers of gaming software and accessories for the new larger, unified market. Unfortunately, though, creating a unified market often ends up causing a standards war. As discussed in more detail in section 6.2.6.1 under the subheading of "Costs of Being Stranded," avoiding a standards war is paramount, since such battles generally delay entry by content providers and users alike until a winning technology has been established.

An alternative means of increasing the potential market size of new technology systems relative to that of older systems is for hardware manufacturers to make newer systems attractive to a wider audience of users than that for older systems. This idea was discussed earlier and illustrated with the examples of mobile phone and Internet devices and of all-in-one printers, copiers, and fax machines, which combine separate user markets into one. Dual-mode technologies can also increase the scope of new technology systems by appealing to users interested in both old and new technology systems or distinct horizontal systems. For example, dual-mode VHS and DVD players attract users from both markets, and dual-mode cell phones (analog/digital or CDMA/W-CDMA/GSM) attract frequent travelers across the United States or between the United States and Europe. Portable devices, such as laptops, PDAs, and iPods, attract users from very mobile, somewhat mobile, and even relatively immobile markets.

> To summarize, hardware manufacturers can encourage content providers to supply content for new technology systems by increasing the relative sizes of markets for newer systems over those for older systems. They can accomplish this by creating single markets for newer systems, either by establishing standards or by making newer systems compatible with existing systems. Alternatively, hardware manufacturers can increase potential market sizes of new systems by increasing their scopes of user appeal relative to those for old systems.

Content Cannibalism and Inaccessibility During the very slow transition from analog to digital television (DTV), the main reason broadcasters have been reluctant to switch their programming from analog to digital is that many viewers do not have TV systems that would enable them to view the digital broadcasts. In other words, digital programming is inaccessible to the large audience of viewers with analog television systems. By providing digital programming, then, broadcasters prevent themselves from being able to generate any revenues on digital content from a large portion of the viewing audience. And the larger the number of viewers with analog systems is, the more reluctant the program broadcasters will be to switch to digital, because they will lose the ability to generate revenues from more viewers.

The case of DTV illustrates how content providers will be reluctant to leave older system markets before their prospects there have been exhausted. Otherwise stated, content providers will be deterred from switching over to supplying content for newer technology systems if sales for new systems come at the expense of or *cannibalize* sales of content for older systems. More specifically, when there are large installed bases of old system users (for example, analog viewers), content suppliers are relatively assured of earning at least minimum amounts of profits by simply continuing to supply content to the existing bases of users. In contrast, if content providers switch their resources over to production of content for new systems (such as DTV), then they lose their ability to generate profits on that new system content from the large base of users of older systems (for example, analog viewers), since these users cannot access the content for newer systems. These lost sales due to inaccessibility or cannibalization issues deter content providers from switching resources out of production of content for older technology systems and into the production of content for new systems. Hardware manufacturers can thus encourage content providers to switch over to providing content for newer technology systems by decreasing demand for old system content, that is, by decreasing the number of old system users or decreasing the amount of content old system users will buy. In section 6.2.4.1, I discussed the methods hardware manufacturers can use to force users to give up on older systems by inducing obsolescence in old system hardware. Briefly, these methods include having hardware manufacturers stop providing support for the old system hardware, for example, by discontinuing the provision of patches, upgrades, parts, or service. Alternatively, they can make all new product and service offerings dependent upon the use of new systems, or in other words incompatible with older systems. Finally, hardware manufacturers can lobby for regulations that effectively make old systems obsolete.

Profitability Because of the evolving capabilities of the technology, users for each generation of telephones have been able to purchase ever more accessories and subscribe to ever more types of services. Current users can enhance the functionality of cell phones, for example, by using various accessories, such as hands-free headsets, belt clips, and extra batteries, and by subscribing to services such as voice mail, text messaging, conference calling, and accessing the Internet. In other words, over time users have been able to get more and more from their telephone systems through enhanced offerings of handsets, services, and accessories. And the relative profitability of serving telephone system users has been increasing in step with the variety of system offerings.

More generally, hardware manufacturers can encourage content providers to switch over to supplying content for new technology systems by making it more profitable for content suppliers to sell to new system users than to old system users. This may be accomplished either by increasing the amount of content each user buys for new systems (such as voice mail, texting, Internet access) or by increasing the profitability of each sale to users for new systems relative to old. Profitability from new system sales of content might be larger for new systems if, for example, hardware manufacturers granted content providers exclusive access to new system users. This is the case for AT&T's provision of services for U.S. users of Apple's iPhone.

Alternatively, the profitability to content providers of serving new system users may be enhanced if their costs decrease faster than the prices users pay for content or services. For example, newer technologies that have enabled telephone service providers to (1) use spectrums more efficiently and (2) provide telephone services over the Internet have decreased the costs per user of providing telephone services.

In the next section we further discuss how hardware manufacturers can help decrease content providers' costs associated with switching to the provision of new system content, which would also help increase the profitability to service providers of moving to new systems.

To summarize, hardware manufacturers can increase the relative profitability to content providers of selling to new technology system users by (1) making new systems more dependent upon the use of content relative to that of older systems, (2) granting content providers exclusive rights to the provision of new system content, and (3) decreasing the relative costs of providing content to new system users.

6.2.6 Switching Costs

In Chapter 1, I discussed different types of costs to users and content providers of switching from older systems to newer ones. During that discussion I also indicated that as networks have become more established, the costs to users and content providers of switching from one system to another have become significant. As such, switching costs can considerably slow the speed of adoption of new technology systems. In this section I discuss how hardware manufacturers can help defray the costs to users and content providers of switching out of older systems and into newer ones, thereby speeding up the pace of new technology adoption.

6.2.6.1 For Users

Switching costs are part of the costs users must incur when adopting new technology systems. To the extent that hardware manufacturers can help users mitigate switching costs, they will decrease the overall costs, thereby increasing the net value of new technology systems to users. Ultimately, this will help speed up the pace of adoption. I now discuss how hardware manufacturers can help mitigate each of the four major types of switching costs previously discussed in section 1.2: (1) costs of uninstalling and disposing of old technology systems and installing new ones, (2) costs of learning how to use new technology systems and getting them ramped up to full productivity, (3) costs of losing network connectivity during transition to new systems, and (4) costs of being stranded in the case that new systems ultimately prove to be less valuable than expected.

Installation Costs In today's interconnected world, it is not always easy to disencumber oneself from the long tentacles of an old technology system, with its associated relationships, service contracts, accessories, product specifications, specially formatted archives, and so on. Although some manufacturers will take trade-ins or otherwise dispose of old models, or help cover the costs of contract termination[14] in order to lower users' costs of switching from one system to another, other technology systems are much more established and require more significant costs of removal. To the extent that uninstall and disposal costs for old systems and/or installation costs for new systems are large, the price hardware manufacturers can charge for new system hardware, or, equivalently, the value users will realize from it, will be constrained by the costs users must incur to remove their old technology systems and install new ones.

There are a couple of techniques hardware manufacturers can employ to help users mitigate the costs of removing old systems and installing new ones. First, hardware manufacturers can wait until product cycles end before introducing new technology systems; that is, they can phase in new systems as service contracts, warranties, product lifecycles, and such for older systems expire. This will decrease, if not eliminate, costs associated with canceling contracts, forgoing services, retrofitting old equipment, and so on.

14. AOL recently received an avalanche of bad publicity when "a Bronx man, who spent 21 minutes trying to cancel his membership, recorded his conversation after an AOL customer service rep refused to cancel his account, despite dozens of requests" (*Reuters* 2006).

Second, hardware manufacturers can make sure they adhere to industry standards. By respecting technical standards or commonly used sizes, specifications, connections, requirements, and so on, hardware manufacturers will decrease any extra costs users might otherwise face in order to readjust their environments to suit the needs of new systems.

Third, hardware manufacturers can anticipate issues or problems users may encounter during the uninstall and reinstall processes and help prepare for them to enable the process to flow as smoothly and as quickly as possible. For example, if special tools are needed to extract old equipment or install new equipment, then hardware manufacturers should either provide access to those tools or at least know where users can find them. If there are regulations surrounding the disposal of old equipment, then hardware manufacturers should be prepared to help users adhere to them. If large-scale file transfer updating from old systems to new ones is required, then hardware manufacturers should make resources available to users to facilitate this process.

> To summarize, hardware manufacturers can minimize users' costs of switching to new systems by (1) waiting until product cycles end before introducing new technology systems, (2) making sure they adhere to industry standards, and (3) being proactive in helping users overcome anticipated problems associated with switching.

Learning and Ramp-up Costs Costs associated with downtime or lost productivity during periods of installation and ramp-up of new technology systems can be significant. The sooner users can achieve full productivity with new systems, the lower the ramp-up costs, and thus the greater the net value, will be of new technology adoption. In fact, SAP has a special program to address this very issue:

> *The SAP® Ramp-Up program is a virtually risk-free way to implement new SAP software as soon as it becomes available. You gain all the advantages of being one of the first companies to adopt a new leading-edge solution—and we make certain you achieve a smooth, cost-effective implementation with an early return on your investment. That's because we supply you with dedicated SAP coaching, support, and knowledge transfer from start to finish.*[15]

15. http://whitepapers.silicon.com/0,39024759,60088020p,00.htm (SAP Customer Services Network Brief, "SAP® Ramp-Up: Getting a Head Start on the Benefits of New SAP Solutions").

To the extent that hardware manufacturers can help users decrease learning costs, downtime, and/or ramp-up costs, they will decrease costs or, equivalently, increase values, and thus speed up the pace of new system adoption. For example, hardware manufacturers can help train new users; they can anticipate problems and provide ready solutions; they can preassemble or modularize components as much as possible, to decrease the time required to assemble or install the new systems; or they can phase in the installation of new systems, so that only small parts of prior systems are inaccessible at any given time during installation or ramp-up.

Losing Network Connectivity during Transition As I discussed earlier, as users become increasingly interconnected with one another within all types of networks, maintaining network connectivity becomes a larger and larger consideration when deciding whether or not to adopt a new technology. In such cases, the transition period during which users leave their old technology systems and lose access to networks, until their entire cohort has adopted newer systems and full access has been reestablished, can be very costly. Anything hardware manufacturers can do to decrease users' lost connectivity while they transition from the old to the new technology systems will increase the value, and thus speed up the pace of adoption, of new systems.

One way for hardware manufacturers to enable users to maintain connectivity to old technology networks while transitioning to new systems is for them to make new technology systems backward-compatible with older systems. With dual-mode (old and new) technology system, users can avail themselves of new system benefits while still being able to "communicate" with old system users.

Alternatively, the hardware manufacturers can make it attractive for connected subgroups to transition together from older to newer technology systems. For example, they can offer package deals or other incentives to all members of a subnetwork who transition to the technology systems at the same time. Family calling plans, which enable all members of a designated group to talk to each other on the phone at reduced rates, provide a good example of this.

Costs of Being Stranded Early adopters of new technology systems face the risk of being stranded if the adoption of new technology systems fails to reach critical mass. This risk tends to delay adoption by many early users until they are assured, or at least face a higher likelihood, that the technology systems will indeed take hold in the marketplace. It follows that hardware manufacturers can speed up the pace of adoption of new technology systems to the extent that they can decrease the risk or costs early adopters face of being stranded.

One of the most decisive ways for hardware manufacturers to decrease the costs to users of either being stranded with the new system or being locked into the wrong system is to have their technology systems established as industry standards or to make their technology systems compatible with other industry standards. Conversely, when hardware manufacturers make their technology incompatible with other similar (horizontal) systems, they create two problems for users. The first problem is that when multiple incompatible technology systems coexist in the market, the size of each of the systems' associated submarkets will be smaller than if the systems were all compatible with one another. Smaller market sizes are less appealing to users, since they entail smaller direct or indirect network effects and/or face critical mass issues; that is, there is a risk that the technology either will not achieve critical mass or will not maintain critical mass, in which case the systems will be forced out of the market, leaving users stranded.

The second problem associated with incompatible technologies is that they may lead to standards or format wars, in which sponsors of incompatible technologies compete to have their technology systems established as actual or de facto industry standards. There are few things that can kill the prospects of early adoption and rapid achievement of critical mass by an otherwise promising new system like having an extended standards or format war. Format wars generally end up causing users and content providers to stand on the sidelines, waiting to see which system will prevail, before they enter the market. Users and content providers that ally themselves with a particular side too early are left with too high a risk that they will end up stranded with the wrong technology.

Examples of standards wars include[16]

- **Vinyl records:** Columbia Records' 12-inch / 33 rpm Long Play (LP) versus RCA Victor's 7-inch / 45 rpm Extended Play (EP)
- **Portable audiotapes:** 8-track cartridges versus compact audiocassettes
- **Home video players:** VHS versus Betamax
- **Personal computers:** Apple versus PC
- **Digital audio data compression:** MP3 versus Ogg Vorbis versus Advanced Audio Coding versus Windows Media Audio

16. http://en.wikipedia.org/wiki/Format_war.

- **Digital video data compression:** Windows Media Video versus RealVideo versus DivX versus QuickTime
- **Memory cards:** CompactFlash versus Memory Stick versus MultiMediaCard / Secure Digital card versus SmartMedia versus XD-Picture Card
- **Satellite radio:** XM Radio versus Sirius Radio
- **High-definition video:** HD-DVD versus Blu-ray

Although it is true that the winners of standards wars stand to generate very large profits, those profits may come with very large costs, such as (1) long delays before standards are decided and users enter the market and (2) risks that entirely new technology systems may come along and usurp the system over which the standards wars are being fought. Rather than participating in standards wars, it may end up being more profitable for manufacturers of new technology systems to simply hook onto the existing or developing standards, acquire installed bases of users, and then perhaps earn their big profits on complementary or evolutionary products or services.

Another method used by software providers in particular to decrease user stranding costs is to place their source code in escrow.

> ***Source code escrow*** *means deposit of the source code of the software into an account held by a third party escrow agent. Escrow is typically requested by a party licensing software (the licensee), to ensure maintenance of the software. The software source code is released to the licensee if the licensor files for bankruptcy or otherwise fails to maintain and update the software as promised in the software license agreement.*[17]

A third way hardware manufacturers can reduce the costs of stranding is to provide systems that contain certain levels of requirements, or that are otherwise complete enough, so that users would still be able to obtain a minimum level of functionality, even if there was no further adoption by users or development or support by providers. An example of this is the service level agreements used by telecommunications companies that guarantee their customers a minimum level of availability over time, with penalties required to be paid when actual availability turns out to be less than that agreed upon (in this case, availability will be too low if there are too many outages).

17. http://en.wikipedia.org/wiki/Source_code_escrow.

To summarize, the most definitive way hardware manufacturers can decrease the risk or costs to early adopters of being stranded, and thus speed up the pace of adoption of new technologies, is by avoiding standards wars or otherwise making their technology systems compatible with existing standards. Other means of decreasing users' stranding costs include placing the technology in escrow so that users will have continued access even if system suppliers go bankrupt, or making technology systems complete enough that they can stand on their own, even if system suppliers do go bankrupt.

6.2.6.2 For Content Providers

Acquiring access to a new market generally requires suppliers to make an investment of time, money, or other resources. More specifically, to gain access to the market of new technology users, content providers may have to pay licensing or other access fees, they may have to buy new equipment or otherwise adapt their facilities to be able to supply new content, and/or they may have to learn new procedures or otherwise gain new knowledge. Any investments content providers are required to make to supply the new system market with content will tend to deter them from switching into production of new system content. It follows that hardware manufacturers can speed up content providers' entry into new system markets by decreasing any of these. For example, hardware manufacturers can decrease licensing or other access fees, perhaps for a limited time. They can also subsidize investments in equipment. And they can facilitate learning requirements by providing training.

As an example, studios have been trying to get movie theaters to upgrade to digital cinema, that is, the use of digital technology to distribute and project movies. However, for this to happen, movie theaters must invest in costly new equipment to be able to advance from film to digital projection:

> *To match or improve the theater experience of movie audiences, a digital cinema system must provide high quality image, sound, subtitles, and captions. Theater managers require server controls for managing and displaying content in multiple theaters, and studios want their content encrypted with secure delivery, playback, and reporting of play times to the distribution company.*[18]

---

18. http://en.wikipedia.org/wiki/Digital_cinema. See also MacBride 2005.

The cost of the new systems, together with the need to coordinate all the requirements, has been a large deterrent to theaters from upgrading. To facilitate the coordination, in March 2002, movie studios and vendors formed a consortium, the Digital Cinema Initiatives (DCI), "to establish a standard architecture for digital cinema."[19] In 2005 the DCI was able to map out a common system that was agreed upon by all and that will finally enable the adoption of digital cinema to move forward:

> *After three years of careful planning, discussion and reaching out to all the various constituencies who make up our industry, DCI member studios are pleased to have reached unanimous agreement on the necessary overall system requirements and specifications for digital cinema ... We now have a unified specification that will allow manufacturers to create products that will be employable at movie theatres throughout the country and, it is hoped, throughout the world. In fact, the specification is being translated for international markets.*
>
> *These specifications should provide a common ground to spur innovation and encourage many more players who were previously resistant to invest capital in technology that may or may not have been viable ... And as the market gets more competitive, the price of the equipment and its installation—previously thought to be a major barrier to digital cinema—will become increasingly affordable, to the point where that stumbling block should no longer be of consequence.*[20]

Another example of how a "manufacturer" can facilitate the supply of content by reducing content providers' costs is the eBay Developers Program, previously discussed in Chapter 1, in which eBay developed a company-sponsored initiative with the purpose of helping third-party developers integrate their businesses with the eBay Platform.

> To summarize, hardware manufacturers can help reduce content providers' switching costs associated with providing new technology system content by decreasing licensing or access fees for new systems, by subsidizing investment costs, and/or by facilitating learning or coordination requirements.

19. http://en.wikipedia.org/wiki/Digital_Cinema_Initiatives.

20. www.dcimovies.com/press/07-27-05.tt2.

6.2.7 Summary of General Responses

In short, hardware manufacturers may be able to encourage the early majority to upgrade to the new technology sooner rather than later by (1) increasing the value of new technology systems relative to that of the old systems—by increasing the value of the technology itself, increasing the size of the installed base, and/or increasing the availability of content; (2) decreasing the price of new technology systems; (3) decreasing the costs to users of switching to new technology systems—by decreasing the costs of uninstalling old systems and installing new ones, decreasing users' costs of learning how to use new technology systems and ramping up to full productivity, decreasing the costs to users of being disconnected from networks, and/or decreasing costs to users of being stranded, should new technology systems not succeed in the marketplace; or (4) inducing obsolescence in old technology systems.

Many of these techniques for speeding up the pace of new technology adoption by the early majority compel hardware manufacturers to accept lower profits than they might otherwise achieve if they do not make such efforts. More specifically, these techniques require hardware manufacturers to incur higher costs, for example, by subsidizing uninstallation of old technology systems or installation of new systems; to generate lower revenues, such as by offering lower prices to some early adopters; or to create technology systems that are less unique than they might otherwise be, such as by making them compatible with other systems. However, the trade-off for sacrificing these profits early on is a greater likelihood that their technology systems will become established in the marketplace. In other words, hardware manufacturers trade lower chances of higher profits for higher chances of lower profits. Yet, if hardware manufacturers act efficiently in trading profits for risk,[21] at the time they begin their journeys, their expected profits should be no less under the higher probability/lower profits scenarios, and risk-averse producers may in fact be better off by having to face less risk of technology adoption.[22]

21. The hardware manufacturer decreases risk efficiently by increasing the reduction of risk until the last dollar spent to increase the chance of attaining critical mass just equals one dollar in forgone profits.

22. It follows from this reasoning that technologies that are more revolutionary in nature, which face a higher chance of not achieving critical mass, must be expected to be commensurately more profitable. Otherwise, it would not be worth the risk to invest in more unique technologies.

6.3 Scenario-Specific Responses to the Chicken-and-Egg Problem

In Chapter 5, I examined four characteristics of new technology systems that impact the eagerness of users to adopt and content providers to supply the new systems: (1) whether potential market sizes (technology values) for new systems are small or large, (2) whether or not direct network effects—that is, user interaction—create significant value for system users, (3) whether or not indirect network effects—that is, the availability of complementary content—create significant value for system users, and (4) whether or not users are sensitive to price.

Then, in previous sections of this chapter, I provided techniques that system innovators might use to attract users and content providers to adopt new technology systems more quickly; that is, these general techniques could apply to new technology systems with any of the market characteristics discussed in Chapter 5. For specific market scenarios, however, certain methods can be particularly effective at enticing new system adoption by users and content providers. I now discuss which techniques are better suited to aid innovators in the adoption of new systems for markets with specific characteristics. A summary of these market-specific techniques is provided in Figure 6-3.

Characteristic	Hardware Manufacture Actions to Speed Adoption by: Users	Content Providers
Market Size		
Small	• Make technology value high • Customization (product versioning) • Early availability of content	• Increase profitability per user (see Increase Relative Value of New System)
Large	• All strategies	• All strategies
User Sensitivity to Price		
High	• Mass production • Penetration pricing • Early availability of content	• Establish expectations of large market potential • Decrease content providers' costs
Low	• Customization (product versioning with price skimming)	• Increase profitability (price skimming) • Customization (product versioning)
Direct Network Effects		
Small	• Make technology value high • Customization (product versioning) • Early availability of content	
Large	• Pentration pricing • Joint adoption by sub-network of users	
Indirect Network Effects		
Small	• Make technology value high • Customization (product versioning) • Penetration pricing	• Establish expectations of large market potential • Decrease content providers' costs
Large	• Early availability of content	• All strategies

Figure 6-3 Scenario-specific responses to speed up the pace of adoption

6.3.1 Market Size

6.3.1.1 Technology Systems for Small Markets

GE spends billions of dollars to develop new systems for generating electricity from nuclear energy. Yet GE has only a small potential market of buyers from which to recover its development costs and generate a profit. It cannot afford to give away nuclear systems to early adopters to get them to validate the systems. Rather, GE must charge relatively high prices to generate large profits from even the first buyers.

What does it take for early adopters in small markets to pay a high price for new technology systems? The first option is for system innovators to offer early adopters new systems with generally high technology values. In GE's case, nuclear power plants must be able to generate much cheaper electricity than coal plants. Another option that can be particularly effective in small markets is for system providers to offer special system customization for each user—adding special features or modifying existing characteristics, to further enhance system value to their customers. In GE's case, it could further enhance the net value of nuclear systems to users by choosing sizes and locations that suit particular users' needs, helping them construct the plants, train the operators, and get them ramped up to full capacity as quickly as possible. GE could further offer warranties or (tailored) maintenance programs to help with glitches encountered by operators during system use.

The second alternative is for system innovators to attract content providers early on so as to create value for early adopters through availability of new system content. Yet when system markets are small, content providers cannot expect to sell to large numbers of users. For small technology markets to be attractive to content providers, then, the expected profits per user must be relatively high. So, when new system markets are small, hardware manufacturers should focus on attracting content providers by enhancing content profitability per user. Possibilities for doing so are discussed in section 6.2.5.

6.3.1.2 Technology Systems for Large Markets

When the market potential for new technology systems is large, hardware manufacturers have more options for speeding up the pace of adoption by users and content providers. System innovators for smaller markets must sell even the first units at a profit, so they must provide the earliest adopters with a lot of value. Innovators for large markets, on the other hand, have more breathing room, and they often lack much of the "do-or-die" sense associated with adoption of systems in smaller markets. In particular, since there are relatively few potential early adopters (and few late adopters) for small market systems, innovators of small market systems have fewer chances to get the ball rolling. If they cannot get their first few target users to adopt their new systems, their window of opportunity might close. With larger market potential, however, system innovators usually have more chances to convince users of the value of new systems. Also, with larger markets, innovators can make up any profits lost on sales to early adopters on sales to the multitudes of later adopters. With this freedom, innovators can, for example, offer early adopters discounts to lure them in, or they can offer special product customization to enhance the appeal of their new systems. They can attract content providers to the sizable profit potential associated with larger markets, or they can attract early users with good reputations. In other words, practically any of the strategies discussed will also apply to markets with large potential size.

6.3.2 User Sensitivity to Price

6.3.2.1 High User Sensitivity to Price

Henry Ford referred to his Ford Model T, which was introduced on October 1, 1908, as "the universal car." Ford specifically engineered his automobile to be low-cost by initiating the assembly line, mass production, and interchangeable parts. "The car was introduced with a price tag of $850 [under $20,000 in 2008 dollars] . . . [and] later sold for as little as $260, without extras, because of production savings Henry Ford passed on to customers."[23]

When users are sensitive to price, systems must be low-priced to get users to adopt. In this case, system innovators have two options especially suited to markets with price-sensitive users. First, for low prices to generate profits, systems must be manufactured at low costs. The most obvious means of achieving low cost of production is through economies of scale. In other words, suppliers can reduce unit costs of manufacturing by automating the production process as much as possible and using large-scale production runs, with little variety in offerings. Of course, large-scale production means that there must be a large number of users. In other words, this technique can be used only for systems with relatively large market potential.

When users are sensitive to price, price discounting will be particularly effective for attracting users. So, in addition to or instead of engineering low-cost systems, innovators of systems with price-sensitive users might offer price discounts to early

23. http://media.ford.com/article_display.cfm?article_id=858.

adopters—that is, use *penetration pricing*—to get them to try out the system and then spread the word, thereby establishing system validation.

Alternatively, when users are sensitive to price, system innovators can attract early users by enhancing the value of the system, that is, giving them more for their money. Since direct networks will be small for early adopters, system innovators are left with the option of enhancing system value through the availability of complementary products and accessories.

As with hardware manufacturers, when users are sensitive to price, content providers must generate profits either by generating large numbers of buyers (that is, on volume) or by generating very low costs per buyer (that is, on moderate margins). In these cases, system innovators will attract content providers to the market either by establishing that the eventual size of the market will be large or by helping to defray content providers' costs of supplying new markets. The preceding sections on general techniques provide more detailed descriptions of how hardware manufacturers can accomplish either of these.

6.3.2.2 Low User Sensitivity to Price

When users are less sensitive to price, they are willing to pay for value. In this case, a doubly effective technique, both to encourage system adoption and to generate higher profits, is for hardware manufacturers and content providers to tailor their offerings to meet specific user needs. This technique is also known as *product versioning* or *price skimming*. Examples of product versioning for hardware manufacturers include charging higher prices for high-definition DVD players, charging higher prices for surround-sound audio systems, charging higher prices for superchargers or other detailing for sports cars, or charging higher prices for the added ability to access systems remotely. Examples of product versioning for content providers include charging higher prices for directors' cut or high-definition DVDs, offering high-octane gasoline, or charging higher prices for premium seating. The prospects of high profits through product versioning and price skimming should attract content providers early on to these new markets.

6.3.3 Direct Network Effects

6.3.3.1 Small Direct Network Effects

When direct network effects are small, the presence of other users does not enhance system values directly. In these cases, attracting early and later adopters requires that new systems have high technology values. Aside from creating systems with generally higher technology values than existing systems, hardware manufacturers can further enhance system values to particular users by customizing them to meet specific user needs. Alternatively, and especially if content availability greatly enhances system values to users, hardware manufacturers can encourage the early provision of complementary products and accessories for their new systems. The preceding sections on general techniques provide more detailed descriptions of how hardware manufacturers can encourage early entry by content providers.

6.3.3.2 Large Direct Network Effects

When direct network effects are large, system innovators will maximize profits by establishing bases of users as quickly as possible so as to generate network effects and enhance system value to later adopters. The clearest strategy for accomplishing this is for system innovators to discount sales of new system hardware to early adopters. Such penetration pricing strategies will entice early adopters to try out new systems, provide validation, and establish bases of users from which later adopters can receive benefits through interaction.

Alternatively, when direct network effects are large, there is an additional effective strategy for encouraging adoption by users. Consider a small group of users who form their own mini network; that is, there is a lot of interaction among members of this small group. For example, these users might form a support group, a gaming group, or a research group, such as ARPA, from which the ARPANET, the precursor of the Internet, was developed. The key is that if large portions of users in these mini networks simultaneously adopt a new system, they will generate a lot of value within the group through direct network effects. In this case, system innovators can create strong momentum by offering packages that encourage joint adoption of new systems by these mini networks of users.

6.3.4 Indirect Network Effects

6.3.4.1 Small Indirect Network Effects

When indirect network effects are small, content availability does not create much system value for users. In this case, users' demand for content will be weak and/or users will not be willing to pay high prices for content. As such, hardware manufacturers should focus on creating and/or enhancing the systems' technology value to users. Customizing hardware to meet specific customer needs will be particularly effective if users are not sensitive to price. If users are sensitive to price, and especially if network effects are strong, then hardware manufacturers can enhance system value to users by encouraging the quick formation of direct network effects, in particular by using penetration pricing.

For content providers, the case of small indirect network effects—that is, weak user demand for content and/or unwillingness to pay high prices—is equivalent to the case in which users are sensitive to prices. As in the case of user sensitivity to price, hardware manufacturers can encourage content suppliers to switch over to the new system (1) by establishing expectations that the new system market will be large—that is, content providers will generate profits on large volume sales; and/or (2) by helping content providers minimize their costs—that is, by helping them squeeze as much margin as possible from sales of content.

6.3.4.2 Large Indirect Network Effects

When indirect network effects are large, content availability greatly enhances system value for users. In these cases, system innovators will most effectively attract users to adopt their new systems by encouraging the early provision of content by third-party suppliers. Furthermore, since users will be eager to buy content in these cases, content providers should be attracted to these new system markets relatively easily. In other words, practically any of the strategies discussed for attracting content providers early on will apply to markets with large indirect network effects.

6.3.5 Summary of Scenario-Specific Responses

In this section I have discussed techniques to aid innovators in the adoption of new systems for markets with specific characteristics. Markets will generally be less inviting to manufacturers of hardware for network systems when market size is small, users are sensitive to price, direct network effects are small, and/or indirect network effects are small. In these cases, hardware manufacturers can help speed up the pace of user adoption of their new systems by making the technology itself (as opposed to its associated networks) as valuable as possible relative to currently available alternatives. Innovators can do this by making the general technology value itself high, by customizing the technology for particular users, or by getting content providers to provide new system content as early as possible. Innovators can help speed the pace of content availability by establishing expectations of large market potential or by increasing the relative profitability of new systems to content providers, by either increasing their revenues or decreasing their costs.

Markets will generally provide greater profit potential to manufacturers of hardware for network systems when market size is large, users are less sensitive to price, direct network effects are large, and/or indirect network effects are large. In these cases, when users are less sensitive to price, hardware manufacturers can help speed up the pace of adoption by both users and content providers through product versioning and price skimming. Hardware manufacturers can further help speed up the pace of adoption by users through penetration pricing when direct network effects are large, or by getting content providers to provide new system content as early as possible.

CHAPTER SEVEN

Summary, Applications, and Extensions

7.1 Key Points from the Analysis

This book examines the holdup problem encountered during the adoption process of a particular class of new technology systems. The systems in this class are those in which (1) both hardware and software components are required to produce the consumption experience for users, (2) the hardware and software components for the technology systems are supplied by providers who act independently from one another, and (3) the technology systems exhibit network effects. I refer to the interactions of these independent suppliers of hardware and software components during the course of introduction of new technology systems as the *Hardware–Software Game.*

In order to understand the dynamics of the Hardware–Software Game, one must first understand the basics of technology system networks and network effects, as well as the nature of technology adoption lifecycles. To this end, in the first part of the book, I provided a review of these basics. During this review, I defined the *total values* that technology systems provide users as the sum of the values of the technologies themselves (their *technology values*) plus the sum of the values of the *direct network effects* (the ability to interact with other users and system validation) and *indirect network effects* (the availability of supplementary content, software, and accessories) associated with the systems. I emphasized that when users adopt new technology systems, they generally use them to replace existing technology systems that accomplish the same or similar tasks, albeit perhaps less effectively or efficiently. Given the existence of established systems, users will adopt new technology systems only if the *extra* value the new systems provide over existing systems exceeds the entire price of the new systems, plus any switching costs. I then cautioned that new system innovators that fail to take into account the fact that users currently use systems that perform similar functions to new systems will overestimate their chances of successful adoption of the new systems.

Next, I discussed the decisions faced by system users regarding adoption of new technology systems and how the impacts of various drivers on users' decisions to adopt change over time. In particular, I noted that (1) larger installed bases will increase user demand for hardware more during the earlier stages of adoption, and the impact will decrease over time as more users adopt; (2) users will tend to be more sensitive to the price of hardware than to the price of content, and their sensitivity will decrease over time as the risks associated with using the systems decrease; and (3) having more content available will increase user demand for content more or less, as later content complements or substitutes for earlier content. At the same time, however, regardless of whether new content complements or substitutes for older content, user demand for content will decrease over time as the market becomes saturated.

I indicated that the problem hardware manufacturers face is to figure out the best way to distribute sales of hardware units across periods so as to maximize total profits. Within the context of that decision, hardware providers will be more successful when (1) they engineer their new systems to have the lowest unit costs possible, (2) their new systems are much better at doing what they are supposed to do than current alternatives, (3) there is a large amount of content available for their new systems as early on as possible, and (4) there is a large installed base of users.

As for content providers, they must figure out the best way to distribute content each period across old and new system users so as to maximize total profits. They will be inclined to transition more quickly to new technology markets when (1) existing markets have been saturated, (2) there is less uncertainty about if and when the new market will become established, (3) the costs of transition to the provision of new system content are lower, and (4) the costs associated with serving separate submarkets of users are lower.

I defined the chicken-and-egg problem so often encountered during the adoption of new technology systems as the delay in adoption due to the simultaneous reluctance of

- Users to adopt new systems before sufficient content becomes available
- Content providers to supply content for new systems before sufficient users have adopted them

I then proposed to use game theory to better understand the structure of the incentives faced by hardware providers, content providers, and users when facing adoption of new technology systems. I noted that the outcome of a *game* is characterized by the fact that

- The payoffs to each player in the game depend on the actions each of the other players take
- Players would be better off if they could coordinate their actions

This is identical to the characterization of the chicken-and-egg problem. By applying game theory to the problem of new technology adoption, I can better understand the dynamics and incentives faced by the various players during the different stages of adoption. I can then use this understanding to modify the market environment to change the incentives of content providers and users to adopt new systems sooner than they would in the absence of the intervention.

I developed a model that simulates the Hardware–Software Game and used this model to undertake a systematic analysis of the structures and incentives of the various players across different types of hardware–software systems and under different market scenarios. My analysis provided three primary insights regarding the incentives of hardware manufacturers (HW). First, the most profitable market environments for HW occur when users are less sensitive to price, when direct network effects are strong, and/or when indirect network effects are strong *when direct network effects are strong.* Second, HW will earn positive profits overall only if there is a lot of content available as early on as possible. And third, when direct network effects are weak, HW will tend to prefer slower rates of adoption. In contrast, when direct network effects are strong, HW will tend to prefer faster rates of adoption, but not too fast.

My analysis provided two primary insights regarding the incentives of content providers (SW). First, as users and content start to move together from the old system to the new one, SW is worse off until a critical mass of users and content has been established for the new system. The corresponding resistance by SW to transition to the new system will be greater when network effects are larger. And second, SW will prefer a very slow pace of new system adoption when serving old system users is profitable, and SW will prefer a very rapid pace of adoption when the new system provides better opportunities than the old one.

In the last part of the book I provided techniques that hardware providers could use to modify the environment so that system users and content providers would prefer to adopt the innovators' new systems sooner rather than later. I started with a discussion of general techniques that HW might employ regardless of the market environment. These methods include (1) basic approaches HW can employ when engineering new systems to enhance the chances of successful adoption, (2) creating expectations of new system adoption, (3) establishing a good reputation, (4) decreasing the relative values of established alternatives to their new systems, (5) increasing the value of their new systems relative to current alternatives, and (6) helping to minimize the switching costs users and content providers might incur when transitioning to HW's new systems. I ended the last chapter with a discussion of scenario-specific techniques hardware manufacturers could use to modify the environment so that system users and content providers would prefer to adopt the innovators' new systems sooner rather than later. In particular, I discussed specific techniques to be used when (1) market sizes of the new systems are small, (2) users are more sensitive versus less sensitive to price, (3) direct network effects are small versus large, or (4) indirect network effects are small versus large.

In their most basic form, the solutions to the chicken-and-egg problem entail making the value proposition of new technology as high as possible for users—that is, making sure the new technology provides substantial benefits over previous generations—while minimizing the costs to users and content providers of switching over to the new technology system. Notable among the switching costs to users is the loss of network connectivity during the transition, which is becoming a more significant issue as users become increasingly connected within their personal, professional, and social networks. Hardware manufacturers can address this issue by making the new technology as compatible and easy to install as possible with users' networks.

Also notable are costs to users and content providers alike of being stranded with a technology that never ends up taking hold. Hardware manufacturers can address this issue by making their hardware compatible with last-generation or same-generation technologies, or by coming to a quick consensus on a single, unified standard or format, to avoid bifurcating the market or having users and content providers wait on the sidelines until hardware manufacturers agree on a format.

7.2 Tools for Applying the Model

7.2.1 Assessing Relative Sizes and Sources of New System Values

Readers can apply the concepts discussed in this book to perform side-by-side comparisons of old and new technology systems. This will help readers better establish the sizes and sources of advantages of new systems over old.

Step 1. Start with a new system, and create a column listing all major features and benefits of that new system, divided into two groups: central features (must haves) and additional features (good to haves). I have created a sample template for the Apple iPhone in Figure 7-1. Next to each feature, indicate whether the source of the benefits associated with that feature is (1) the technology itself (tech), (2) direct network effects (DNE), or (3) indirect network effects (INE).

Step 2. Next, create adjacent columns listing all existing technologies that users must combine to be able to access all the central features of the new system. In the iPhone example, assuming that playing music is a central feature, users would have to use both a PDA, such as a BlackBerry or Palm Pilot, together with a portable music player, such as an iPod, to acquire all the central features provided by the iPhone. Add to the list of features any central or additional features provided by the old system(s) that the new system does not provide.

Step 3. Now, under the list of features, list each of the components users must purchase, together with their prices, to be able to enjoy the consumption experience provided by the various technology systems.

Step 4. Finally, under the list of system prices, list any switching costs users must incur to switch from the existing system(s) to the new one. In the iPhone example, I assume users break their service contracts four months early, resulting in a $200 fee. I also assume users can resell their PDAs and iPods, say, on eBay, for a total of $125. Finally, I assume it takes users about two hours to learn how to use the iPhone, and they value their time at $25 per hour.

Feature [A]	Source (Tech, DNE, INE) [B]	New System iPhone [C]	Old System PDA (BlackBerry, Palm) [D]	Portable Music Player (iPod) [E]
Step 1 / Step 2				
Must Have				
[1] Cell Phone	INE, DNE	X	X	
[2] Organizer	Tech	X	X	
[3] Internet Access	INE	X	X	X
[4] Access e-mail	INE, DNE	X	X	
[5] Play Music	INE	X		X
Good to Have				
[6] Sync with Computer	Tech	X	X	X
[7] Take Pictures	Tech	X		
[8] Applications	INE	X	X	X
Step 3				
System Price				
[9] Handset		$200	$300	$250
[10] Connection Services (2 Years)		$1,920	$1,200	
Step 4				
Switching Costs				
[11] Break Service Contract		$200		
[12] Dispose of Old HW (Resell)		($125)		
[13] Learn to Use iPhone		$50		
Total Costs		$2,245	$1,500	$250

Figure 7-1 Side-by-side comparison: new versus old systems

The completed table provides an overview of the distribution of features across old and new systems. In particular, the table provides information about whether the features can be immediately accessed by users (they rely on the technology itself), they rely on availability of content (INE), or they rely on the presence of a large installed base of users (DNE). If the features of the new system rely heavily on the availability of content, then system innovators should try to induce content providers to provide that content early on. Similarly, if the features of the new system rely heavily on direct network effects, then system innovators should either make their system networks compatible with existing networks or try to induce as many new users as possible to adopt early on. If the price plus switching costs associated with the new system are high, then users will not adopt unless the major features the new system offers that the current systems do not are sufficiently valuable. Finally, if switching costs are high, system innovators should take action to help adopters mitigate those costs.

7.2.2 Better Understanding User Demand for New System Hardware

This section provides readers who are interested in better understanding user demand for new system hardware with two types of exercises they can apply to technology systems of their choosing. The first exercise forces readers to think about technology markets, together with their associated networks of users and content, in more exacting terms. In the second exercise, users apply their understanding of technology markets to estimate user demand for new system hardware.

7.2.2.1 Defining Technology System Markets

When modeling user demand for technology systems, there are two important issues the system innovator must consider:

1. Who are the potential adopters of the system?
2. Who should be included in the base of installed users?

How these questions are answered will significantly affect the size of network effects that the system exhibits. As an exercise, readers can apply the concepts discussed in the book to define alternative groupings of markets and associated sizes of network effects for their own technology systems.

As a specific example, return to the example of the iPhone discussed in the previous section. Consider various groups of technology system users:

- Users of smartphones
- Users of iPods
- Users of cell phones
- Users of PDAs
- Users of PCs/laptops
- Users of telephones

To a greater or lesser extent, users of each of these technology systems do interact or could interact with users of iPhones, and users of each of these systems share some types of content with users of iPhones. Furthermore, compared with nonadopters, users of each of these systems will incur lower switching costs in terms of learning how to use the iPhone, since they can transfer knowledge they have already gained from using their current systems. Finally, users of these systems can also serve to validate, to a greater or lesser extent, the aspects of the iPhone with which they are already familiar. Clearly, though, user demand for hardware will be more or less sensitive to both the size of the "installed base" and to the "availability of content," depending on which of these users of different interlinking technology systems is included in the counts; that is, the size of network effects for iPhones will be larger or smaller, depending on which of the users are included in the definition of iPhone system users.

7.2.2.2 Estimating User Demand for Technology System Hardware

System innovators can apply the example described in the book in section 3.1.4.2 and illustrated in Figures 3-4 and 3-5, as applied to equation (C-1) in Appendix A, to their own technology systems. This exercise will help system innovators estimate the prices they will be able to charge and/or the amount of content availability that users will require before adoption will reach critical mass.

7.2.3 Better Understanding Optimal Allocations of Content

Content providers can apply the example described in section 3.2.3 and illustrated in Figures 3-9 and 3-10, as applied to equation (C-2) in Appendix A, to their own technology systems. This exercise will help content providers determine if and when to switch their resources over to providing content for new systems and/or when to raise or lower their prices.

7.3 Extensions of the Analysis

7.3.1 Estimates of Network Effect Elasticities

The concept of price elasticity (user sensitivity to price) is well known, and there are numerous studies in which researchers empirically estimate price elasticities for various products.[1] However, the concepts of direct and indirect network effect elasticities (user sensitivities to the sizes of installed bases and availability of content) are less well known. Furthermore, the results of my simulation model are relatively sensitive to large changes in the size of network effect elasticities. To my knowledge, there are no empirical studies that estimate network effect elasticities for different types of network systems. Having such estimates would help system suppliers better optimize their entry decisions.

7.3.2 Network System Routes to Market

The Hardware–Software Game is premised on the independence of hardware manufacturers and content providers. This holds true for such technology systems as DVD players and discs, and gaming consoles and cartridges. However, end users access a significant portion of hardware–software technology systems in ways other than purchasing components separately from independent hardware and software suppliers. In particular, sometimes one firm jointly supplies hardware and software components to end users, as in the case of Gillette razors and blades or Hewlett-Packard printers and ink. In other cases, the hardware components are sold bundled with software or content, for example, in the case of cellular phones and cellular service, or cable set-top boxes and broadcast services. Alternatively, software components can be sold bundled with hardware, for example, in the case of PCs.

The existence of these different routes to market then begs three questions regarding optimal actions of hardware and software suppliers:

1. When will hardware and software providers be led to integrate (through contracts, joint ventures, or mergers) so as to be able to bundle their products and sell them jointly, as opposed to selling them independently from one another to end users?
2. When will we see software providers as the distributors of system components to end users, where they sell their software bundled with hardware (as in the case of cable television services and cellular phone services)?

1. See, for example, Bayus 1992, Parker 1992, and Parker and Neelamegham 1997.

3. When will we see hardware providers as the distributors of system components to end users, where they sell their hardware bundled with software (as in the case of PCs)?

The answers to these questions lie in the consideration of three factors.

7.3.2.1 Factor 1: End-User Valuations Differ across Components

End users differ in their relative valuations of the various hardware and software/content components that combine to generate the consumption experience. Some end users will pay high prices for high-quality hardware that has all the bells and whistles, whereas other users are interested only in the bare-necessity hardware. Some end users will pay a high price for high quality or large volumes or varieties of content, and other end users will pay for only the essential content. Furthermore, the end users who are willing to pay high prices for hardware are not necessarily the same end users who are willing to pay high prices for large volumes or varieties of content.

7.3.2.2 Factor 2: Bundling of Components Increases Profits

Suppliers of the various technology components will generally be able to generate greater total revenues when they bundle the components and sell them jointly in packages, as opposed to selling the hardware and software components separately.

7.3.2.3 Factor 3: Dominant versus Subordinate Suppliers

Some technology systems involve one or a few *dominant* components with many *subordinate* components. I define a *dominant* component as one for which the suppliers are relatively few in number, whereas *subordinate* component suppliers are many and diffuse. For example, in the case of portable music players (Sony Walkmans or Apple iPods), there are a few suppliers of the hardware components for the system. In contrast, the content components—songs, lectures, books, and so on—are supplied by a proliferation of different providers.

Alternatively, *dominant* suppliers can be ones that have ongoing relationships with end users, whereas *subordinate* suppliers have one-time contacts with users. Examples of these types of dominant suppliers are suppliers of cellular phone and television (cable or satellite) programming services. The use of cell phones and television programming services entail the one-time provision of hardware (handsets and set-top boxes), with ongoing subscriptions for connection and programming services.

Technologies that conform to the models in which there are dominant suppliers of one system component and subordinate suppliers of the other system components lend themselves to having the dominant component suppliers provide all hardware and software components to end users, whereas technology systems for which there are no dominant suppliers do not.

Further analysis of routes to market is a natural extension of the analysis of the Hardware–Software Game. In particular, the number of suppliers of the various components comprising technology systems has exploded with the globalization of markets. What has been the market response to these dynamics? When have suppliers integrated with one another to overcome the holdup problem? Is hardware becoming more subservient to software, or vice versa?

APPENDIX A

Model of the Hardware–Software Game

A.1 Definition of Key Terms

A.1.1 Technology User Dynamics

Assume that there are $\bar{N}$ users of an existing technology system when a new evolutionary version of the system is introduced into the market. Once the new system is introduced into the market, a fraction, n_t, of the $\bar{N}$ people who currently own the old system will upgrade to the new system during period t, where $0 \le n_t \le 1$. In any given period t, the number of people who upgrade from the old to the new system—that is, the flow of new system users, N_t—is equal to the portion of users who upgrade during the period, n_t, times the total number of users, $\bar{N}$, so that $N_t = n_t\bar{N}$. The stock of new system users during the period, $\hat{N}_t$, corresponding to the flow of users described above, N_t, is the fraction of users who have upgraded to the new system by the end of the period, $\hat{n}_t = n_1 + n_2 + ... + n_t$, times the total number of technology users, $\bar{N}$, so that $\hat{N}_t = \hat{n}_t\bar{N}$.

A.1.2 Technology Content Dynamics

The content provider has the resources to produce a fixed amount of content each period, $\bar{C}$, and must decide how to allocate that content between the old and the new technology systems. For any given period t, the amount of content that providers generate for the old technology, Z_t, plus the amount of content that providers generate for the new technology, D_t, is limited to the fixed amount of output, $\bar{C}$:

$$\bar{C} = Z_t + D_t \qquad \text{for all t}$$

During the period of introduction of the new technology system, the content provider will divide its resources, devoting a portion $(1-d_1) = \frac{Z_1}{\bar{C}}$ to producing an amount Z_1 of content for the old system and devoting a portion $d_1 = \frac{D_1}{\overline{C}}$ of its resources to producing an amount D_1 of content for the new system. I assume that once content providers have decided to allocate a certain portion of content to the production of new technology content, they will never choose to dedicate less than that portion in later periods; that is, I assume $0 \le d_1 \le d_2 \le ... \le 1$.

The two sets of parameters—the fraction of users who have switched from the old to the new technology system each period, $\{n_1, n_2, ..., n_t\}$, and the fraction of content that has been switched from the old to the new system each period, $\{d_1, d_2, ..., d_t\}$—are the key parameters in characterizing the dynamics of the transition process. They tell us how fast users have upgraded their hardware to the newest generation and how fast content providers have switched over from providing content for the old system to providing content for the new system.

Variable	Description
$\overline{N}$	Total number of old and new technology system users
n_t	Fraction of new system users added during period t (flow measure): $0 \le n_t \le 1$
N_t	Number of new system users added during period t (flow measure): $N_t = n_t\overline{N}$
$\hat{n}_t$	Fraction of users at the end of time t who have upgraded to the new system: $\hat{n}_t = n_1 + n_2, ... + n_t$
$\hat{N}_t$	Total users at the end of time t who have upgraded to the new system (stock measure): $\hat{N}_t = \hat{n}_t\overline{N}$
$\overline{C}$	Total amount of content produced each period (flow measure)
D_t	Amount of new system content produced during period t (flow measure): $D_t = n_t\overline{C}$
d_t	Fraction of new system content produced during period t (flow measure): $0 \le d_1 \le d_2 \le ... \le 1$

Figure A-1. Summary of quantity notations for the Hardware–Software Game

A.2 User Demand Functions

A.2.1 User Demand for New Technology Hardware

The user demand function for hardware may exhibit direct and/or indirect network effects. If there are direct network effects, then the demand for the new technology hardware during period t will be greater when the number of people who have bought the new hardware during previous periods is greater, that is, when $\hat{N}_{t-1} = \hat{n}_{t-1}\overline{N}$ (the installed base) is greater. If there are indirect network effects, then the demand for new technology hardware during period t will be greater when there is more content available for the new hardware, that is, when $D_t = d_t\bar{C}$ is larger. User demand for new system hardware during period t will also be higher when the price of new system hardware, P_{D_t}, and the price of software, r_{D_t}, are both lower. Finally, user demand for new system hardware is based on the marginal value provided by the new technology system over the old technology system. So we have demand for the new system hardware as

$$n_t\bar{N} = \left\{a_t \bullet \left(d_t\bar{C}\right)^{\alpha_t} \bullet \left(\hat{n}_{t-1}\bar{N}\right)^{\theta_t} \bullet \left(P_{D_t} + r_{D_t}\right)^{-\varepsilon_t}\right\} - \left\{a'_t \bullet \left[\left(1-d_t\right)\bar{C}\right]^{\alpha'_t} \bullet \left[\left(1-\hat{n}_{t-1}\right)\bar{N}\right]^{\theta'_t} \bullet r_{Z_t}^{-\varepsilon'_t}\right\} \quad \text{(C-1)}$$

where

a_t is the constant of demand for new system hardware

$d_t\bar{C}_t$ is the amount of content (software and accessories) available for the new technology hardware at the end of time t

α_t is the elasticity of the demand for new technology hardware with respect to changes in the amount of software that is available for the new technology, that is, the magnitude of indirect network effects

$\hat{n}_{t-1}\bar{N}$ is the stock of users who have upgraded to the new technology as of the end of the previous period, $t-1$

θ_t is the elasticity of the demand for new technology hardware with respect to changes in the number of people who have already adopted the new technology, that is, the magnitude of direct network effects

P_{D_t} is the price of the new technology hardware

r_{D_t} is the price of new technology content

ε_t is the elasticity of demand for new hardware and content with respect to changes in price

a'_t is the constant of demand for old system hardware

α'_t is the elasticity of the demand for old technology hardware with respect to changes in the amount of software that is available for the old technology

θ'_t is the elasticity of the demand for old technology hardware with respect to changes in the number of people who have not yet adopted the new technology

r_{Z_t} is the price of old technology content

ε'_t is the elasticity of demand for old hardware and content with respect to changes in price

A.2.2 User Demand for New Technology Content

Since only people with the new technology hardware will buy software for the new technology system, user demand for new content, $D_t = d_t\bar{C}$, is higher when there are more people with the new technology, that is, when $\hat{n}_t\bar{N}$ is higher. Demand for new content will also be greater when there is more new content available, that is, when $d_t\bar{C}$ is higher. Finally, user demand for new content will be higher when the price of new content, r_{D_t}, is lower:

$$D_t = d_t\bar{C} = b_t \bullet \left(d_t\bar{C}\right)^{\varphi_t} \bullet \left(\hat{n}_t\bar{N}\right)^{\gamma_t} \bullet r_{D_t}^{-\kappa_t} \qquad \textbf{(C-2)}$$

where

b_t is the constant of demand

φ_t is the elasticity of consumer demand for content with respect to the availability of content

γ_t is the elasticity of consumer demand for content with respect to the size of the installed base

κ_t is the elasticity of consumer demand for content with respect to price

When users are deciding whether or not to upgrade to a particular hardware–software system, they tend to have choices among various alternatives, be they similar technologies or alternative solutions altogether. At this point, then, given the alternatives, users tend to be more sensitive to the price of the hardware and software components. After users commit to a particular system and buy the hardware, they no longer have the same content choices they did before they decided on that particular system. Rather, they are limited to the content that is compatible with the particular system they chose. As a result, user sensitivity to the price of hardware and software tends to be higher before users buy a new technology than after they buy the new system. In my model, this means that the exponents on price in the hardware demand function in equation (C-1), ε, will tend to be higher than the exponents on price in the software demand function in equation (C-2), κ.

Users who have not upgraded to the new technology will continue to purchase content for the old technology. User demand for old content, $Z_t = (1-d_t)\bullet\bar{C}$, is higher when there are more people with the old technology—that is, $(1-\hat{n}_t)\bullet\bar{N}$ is higher—and when the availability of old content, $(1-d_t)\bullet\bar{C}$, is greater. User demand for old content will also be higher when the price of old content, r_{Z_t}, is lower:

$$Z_t = (1-d_t)\bullet\bar{C} = z_t\bullet\left[(1-d_t)\bullet\bar{C}\right]^{\varphi'_t}\bullet\left[(1-\hat{n}_t)\bullet\bar{N}\right]^{\gamma'_t}\bullet r_{Z_t}^{-\kappa'_t} \quad \text{(C-3)}$$

where

z_t is the demand constant

φ'_t is the elasticity of user demand for old technology content with respect to changes in the supply of old technology content

γ'_t is the elasticity of user demand for old technology content with respect to changes in the installed base of users with the old technology

κ'_t is the elasticity of user demand for content with respect to changes in price

Variable	Description
θ_t	Elasticity of user demand for hardware with respect to changes in the installed base of users during period t
α_t	Elasticity of user demand for hardware with respect to changes in the supply of content during period t
ε_t	Elasticity of user demand for hardware with respect to changes in the price of content during period t
γ_t	Elasticity of user demand for content with respect to changes in the installed base of users during period t
φ_t	Elasticity of user demand for content with respect to changes in the supply of content during period t
κ_t	Elasticity of user demand for content with respect to changes in the price of content during period t

Figure A-2. Summary of elasticity notations for the Hardware–Software Game

A.3 Provision of Hardware

I assume that new technology hardware is provided by a single (representative) manufacturer. The hardware manufacturer is a profit seeker. Its profits will be higher if it sells more new technology hardware, if the price for each unit of new technology hardware it sells is higher, and/or if the cost of manufacturing new technology hardware is lower.

Learning by doing, technology advances, and economies of scale and specialization cause the per-unit costs of manufacturing new technology hardware to drop over time, as the total number of units of new technology hardware sold increases. Profits of the hardware manufacturer during time t, Π_t^{HW}, are thus defined as the price P_t less cost $c \bullet \left(1-\sum_{j=1}^{t-1} n_j\right)$ of each unit of new technology hardware sold, times the number of new technology hardware units sold during the period $N_t = n_t\bar{N}$:

$$\Pi_t^{HW} = \left[P_{D_t} - c \bullet \left(1-\sum_{j=1}^{t-1} n_j\right)\right] \bullet \left(n_t\bar{N}\right) \qquad \textbf{(HW-1)}$$

The cost function, $c \bullet \left(1-\sum_{j=1}^{t-1} n_j\right)$, reflects the fact that the per-unit costs of manufacturing new technology hardware start at c, then drop toward zero over time as the stock of new technology hardware units sold during previous periods, $\sum_{j=1}^{t-1} n_j$, increases.

For any given period, the hardware manufacturer's problem is to set the price of new technology hardware, P_t, so as to sell new technology hardware to a portion, n_t, of users, taking as given the user demand functions, (C-1) through (C-3), and the amount of new system content offered by the content provider, $D_t = d_t \bullet \bar{C}$. More specifically, the hardware manufacturer's problem is:

Given d_t, choose n_t to MAX

$$\Pi^{HW} = \sum_{t=1}^{\infty} \delta^{t-1} \left[P_{D_t} - c \bullet \left(1-\sum_{j=1}^{t-1} n_j\right)\right] \bullet \left(n_t\bar{N}\right) \qquad \textbf{(HW-2)}$$

subject to the following:

User demand functions for hardware:

$$\begin{aligned} n_t\bar{N} = & \left\{a_t \bullet \left(d_t\bar{C}\right)^{\alpha_t} \bullet \left(\hat{n}_{t-1}\bar{N}\right)^{\theta_t} \bullet \left(P_{D_t} + r_{D_t}\right)^{-\varepsilon_t}\right\} \\ & - \left\{a_t' \bullet \left[\left(1-d_t\right)\bar{C}\right]^{\alpha_t'} \bullet \left[\left(1-\hat{n}_{t-1}\right)\bar{N}\right]^{\theta_t'} \bullet r_{Z_t}^{-\varepsilon_t'}\right\} \end{aligned} \qquad \textbf{(C-1)}$$

User demand functions for software:

$$D_t = d_t\bar{C} = b_t \bullet \left(d_t\bar{C}\right)^{\varphi_t} \bullet \left(\hat{n}_t\bar{N}\right)^{\gamma_t} \bullet r_{D_t}^{-\kappa_t} \qquad \textbf{(C-2)}$$

$$Z_t = \left(1-d_t\right) \bullet \bar{C} = z_t \bullet \left[\left(1-d_t\right) \bullet \bar{C}\right]^{\varphi'_t} \bullet \left[\left(1-\hat{n}_t\right) \bullet \bar{N}\right]^{\gamma'_t} \bullet r_{Z_t}^{-\kappa'_t} \qquad \textbf{(C-3)}$$

Restrictions on transition variables:

$$n_1 + n_2 + \ldots \leq 1 \qquad \textbf{(T-1)}$$

$$0 \leq d_1 \leq d_2 \leq \ldots \leq 1 \qquad \textbf{(T-2)}$$

A.4 Provision of Content

I assume that costs per unit of content are independent of the amount of content produced and that there is a one-time fixed cost, F_D, during the period when the content provider first begins supplying content for the new technology. These fixed costs account for any new access fees, equipment, training, and so on in which the content provider must invest in order to be able to produce new system content.

The content provider's revenues come from users who purchase content, where users with the old technology system buy content for the old system, and users with the new technology system buy content for the new system. (I assume that once consumers upgrade to the new technology system, backward compatibility of the new system means they can still use their inventory of content for the old system, but as soon as they upgrade to the new system, any new content that they buy is content for the new system.) The profit function of the content provider in the Hardware–Software Game is then the sum of profits for sales of new technology content, D_t, plus profits for sales of old technology content, Z_t:

$$\begin{aligned}\Pi_t^{SW} &= \left(r_{D_t} - c_{D_t}\right) \bullet D_t - F_D + \left(r_{Z_t} - c_{Z_t}\right) \bullet Z_t \\ &= \left(r_{D_t} - c_{D_t}\right) \bullet d_t\bar{C} - F_D + \left(r_{Z_t} - c_{Z_t}\right) \bullet \left(1-d_t\right) \bullet \bar{C} \\ &= \left[\left(r_{D_t} - c_{D_t}\right) - \left(r_{Z_t} - c_{Z_t}\right)\right] \bullet d_t\bar{C} - F_D + \left(r_{Z_t} - c_{Z_t}\right) \bullet \bar{C}\end{aligned} \qquad \textbf{(SW-1)}$$

where resource constraints limit the total amount of output, $Z_t + D_t = \bar{C}$, and the fixed cost of investment required to produce the new technology content, F_D, is taken only in the first period in which the content provider starts providing content for the new technology, that is, in the first period t for which $D_t > 0$.

The content provider's problem is:

Given n_t, choose d_t to MAX

$$\Pi^{SW} = \sum_{t=1}^{\infty} \delta^{t-1}\left[\left(r_{D_t} - c_{D_t}\right)\bullet d_t\bar{C} - F_D + \left(r_{Z_t} - c_{Z_t}\right)\bullet\left(1-d_t\right)\bullet\bar{C}\right] \quad \textbf{(SW-2)}$$

subject to the following:

User demand functions for hardware:

$$\begin{aligned} n_t\bar{N} = &\left\{a_t\bullet\left(d_t\bar{C}\right)^{\alpha_t}\bullet\left(\hat{n}_{t-1}\bar{N}\right)^{\theta_t}\bullet\left(P_{D_t}+r_{D_t}\right)^{-\varepsilon_t}\right\} \\ &-\left\{a'_t\bullet\left[\left(1-d_t\right)\bar{C}\right]^{\alpha'_t}\bullet\left[\left(1-\hat{n}_{t-1}\right)\bar{N}\right]^{\theta'_t}\bullet r_{Z_t}^{-\varepsilon'_t}\right\} \end{aligned} \quad \textbf{(C-1)}$$

User demand functions for software:

$$D_t = d_t\bar{C} = b_t\bullet\left(d_t\bar{C}\right)^{\varphi_t}\bullet\left(\hat{n}_t\bar{N}\right)^{\gamma_t}\bullet r_{D_t}^{-\kappa_t} \quad \textbf{(C-2)}$$

$$Z_t = \left(1-d_t\right)\bullet\bar{C} = z_t\bullet\left[\left(1-d_t\right)\bullet\bar{C}\right]^{\varphi'_t}\bullet\left[\left(1-\hat{n}_t\right)\bullet\bar{N}\right]^{\gamma'_t}\bullet r_{Z_t}^{-\kappa'_t} \quad \textbf{(C-3)}$$

Restrictions on transition variables:

$$n_1 + n_2 + \ldots \le 1 \quad \textbf{(T-1)}$$

$$0 \le d_1 \le d_2 \le \ldots \le 1 \quad \textbf{(T-2)}$$

A.5 Three-Period Model

Suppose we were to consider a three-period model of the Hardware–Software Game. In this case:

- **Period 0** represents the pre-new-technology era, during which all users have the old technology and all content is provided for the old technology, $n_0 = 0$ and $d_0 = 0$.
- **Period 1** is the early transitional period, during which innovators and early adopters upgrade to the new technology system and the content provider makes some content for the new system, while the remaining users and content remain with the old system, $0 \le n_1 \le 1$ and $0 \le d_1 \le 1$.
- **Period 2** is the later transitional period, during which the early and possibly late majority upgrades to the new technology and the content provider makes some content for the new system, while the remaining users and content remain with the old system, $0 \le n_2 \le 1$ and $0 \le d_2 \le 1$.
- **Period 3** represents the new technology era when all users have upgraded to the new technology and all content is provided for the new technology, $\hat{n}_3 = 1$ and $d_3 = 1$.

A.5.1 Derivation of the Hardware Manufacturer's Problem

The hardware manufacturer's problem in the three-period game is:

Given n_1, n_2, and n_3, choose d_1, d_2, and d_3 to MAX

$$\begin{aligned}\Pi^{HW} = &\left\{\left[P_{D_1} - c\right] \bullet n_1\bar{N}\right\} + \left\{\delta \bullet \left[P_{D_2} - c \bullet (1 - n_1)\right] \bullet n_2\bar{N}\right\} \\ &+ \left\{\delta^2 \bullet \left[P_{D_3} - c \bullet (1 - n_1 - n_2)\right] \bullet (1 - n_1 - n_2)\bar{N}\right\}\end{aligned} \qquad \textbf{(HW-3)}$$

subject to the following:

First-, second-, and third-period user demand functions for hardware:

$$n_1\bar{N}=\left\{a_1\cdot\left(d_1\bar{C}\right)^{\alpha_1}\cdot\left(P_{D_1}+r_{D_1}\right)^{-\varepsilon_1}\right\} -\left\{a_1'\cdot\left[\left(1-d_1\right)\bar{C}\right]^{\alpha_1'}\cdot\bar{N}^{\theta_1'}\cdot r_{Z_1}^{-\varepsilon_1'}\right\} \tag{C-1A}$$

$$n_2\bar{N}=\left\{a_2\cdot\left(d_2\bar{C}\right)^{\alpha_2}\cdot\left(n_1\bar{N}\right)^{\theta_2}\cdot\left(P_{D_2}+r_{D_2}\right)^{-\varepsilon_2}\right\} -\left\{a_2'\cdot\left[\left(1-d_2\right)\bar{C}\right]^{\alpha_2'}\cdot\left[\left(1-n_1\right)\bar{N}\right]^{\theta_2'}\cdot r_{Z_2}^{-\varepsilon_2'}\right\} \tag{C-1B}$$

$$\left(1-n_1-n_2\right)\bar{N}=\left\{a_3\cdot\bar{C}^{\alpha_3}\cdot\left[\left(n_1+n_2\right)\bar{N}\right]^{\theta_3}\cdot\left(P_{D_3}+r_{D_3}\right)^{-\varepsilon_3}\right\} \tag{C-1C}$$

First-, second-, and third-period user demand functions for new system software:

$$D_1=d_1\bar{C}=b_1\cdot\left(d_1\bar{C}\right)^{\varphi_1}\cdot\left(n_1\bar{N}\right)^{\gamma_1}\cdot r_{D_1}^{-\kappa_1} \tag{C-2A}$$

$$D_2=d_2\bar{C}=b_2\cdot\left(d_2\bar{C}\right)^{\varphi_2}\cdot\left[\left(n_1+n_2\right)\bar{N}\right]^{\gamma_2}\cdot r_{D_2}^{-\kappa_2} \tag{C-2B}$$

$$D_3=d_3\bar{C}=b_3\cdot\left(d_3\bar{C}\right)^{\varphi_3}\cdot\bar{N}^{\gamma_3}\cdot r_{D_3}^{-\kappa_3} \tag{C-2C}$$

First- and second-period user demand functions for old system software:

$$Z_1=\left(1-d_1\right)\cdot\bar{C}=z_1\cdot\left[\left(1-d_1\right)\cdot\bar{C}\right]^{\varphi_1'}\cdot\left[\left(1-n_1\right)\bar{N}\right]^{\gamma_1'}\cdot r_{Z_1}^{-\kappa_1'} \tag{C-3A}$$

$$Z_2=\left(1-d_2\right)\cdot\bar{C}=z_2\cdot\left[\left(1-d_2\right)\cdot\bar{C}\right]^{\varphi_2'}\cdot\left[\left(1-n_1-n_2\right)\cdot\bar{N}\right]^{\gamma_2'}\cdot r_{Z_2}^{-\kappa_2'} \tag{C-3B}$$

Restrictions on transition variables:

$$n_1+n_2+n_3=1 \tag{T-1A}$$

$$0\le d_1\le d_2\le d_3=1 \tag{T-2A}$$

Solving and substituting, we get

$$D_1 = d_1\bar{C} = b_1 \bullet (d_1\bar{C})^{\varphi_1} \bullet (n_1\bar{N})^{\gamma_1} \bullet r_{D_1}^{-\kappa_1}$$

$$r_{D_1}^{-\kappa_1} = \left[\frac{d_1\bar{C}}{b_1 \bullet (d_1\bar{C})^{\varphi_1} \bullet (n_1\bar{N})^{\gamma_1}}\right] \qquad \textbf{(C-2A')}$$

$$r_{D_1} = \left[b_1 \bullet (d_1\bar{C})^{\varphi_1 - 1} \bullet (n_1\bar{N})^{\gamma_1}\right]^{\frac{1}{\kappa_1}}$$

$$D_2 = d_2\bar{C} = b_2 \bullet (d_2\bar{C})^{\varphi_2} \bullet \left[(n_1 + n_2)\bar{N}\right]^{\gamma_2} \bullet r_{D_2}^{-\kappa_2}$$

$$r_{D_2}^{-\kappa_2} = \left\{\frac{d_2\bar{C}}{b_2 \bullet (d_2\bar{C})^{\varphi_2} \bullet \left[(n_1 + n_2)\bar{N}\right]^{\gamma_2}}\right\} \qquad \textbf{(C-2B')}$$

$$r_{D_2} = \left\{b_2 \bullet (d_2\bar{C})^{\varphi_2 - 1} \bullet \left[(n_1 + n_2)\bar{N}\right]^{\gamma_2}\right\}^{\frac{1}{\kappa_2}}$$

$$D_3 = \bar{C} = b_3 \bullet \bar{C}^{\varphi_3} \bullet \bar{N}^{\gamma_3} \bullet r_{D_3}^{-\kappa_3}$$

$$r_{D_3}^{-\kappa_3} = \left[\frac{\bar{C}}{b_3 \bullet \bar{C}^{\varphi_3} \bullet \bar{N}^{\gamma_3}}\right] \qquad \textbf{(C-2C')}$$

$$r_{D_3} = \left(b_3 \bullet \bar{C}^{\varphi_3 - 1} \bullet \bar{N}^{\gamma_3}\right)^{\frac{1}{\kappa_3}}$$

$$Z_1 = (1 - d_1) \bullet \bar{C} = z_1 \bullet \left[(1 - d_1) \bullet \bar{C}\right]^{\varphi_1'} \bullet \left[(1 - n_1)\bar{N}\right]^{\gamma_1'} \bullet r_{Z_1}^{-\kappa_1'}$$

$$r_{Z_1}^{-\kappa_1'} = \frac{(1 - d_1) \bullet \bar{C}}{z_1 \bullet \left[(1 - d_1) \bullet \bar{C}\right]^{\varphi_1'} \bullet \left[(1 - n_1)\bar{N}\right]^{\gamma_1'}} \qquad \textbf{(C-3A')}$$

$$r_{Z_1} = \left\{z_1 \bullet \left[(1 - d_1) \bullet \bar{C}\right]^{\varphi_1' - 1} \bullet \left[(1 - n_1)\bar{N}\right]^{\gamma_1'}\right\}^{\frac{1}{\kappa_1'}}$$

$$Z_2 = (1-d_2)\bullet\bar{C} = z_2\bullet\left[(1-d_2)\bullet\bar{C}\right]^{\varphi_2'}\bullet\left[(1-n_1-n_2)\bullet\bar{N}\right]^{\gamma_2'}\bullet r_{Z_2}^{-\kappa_2'}$$

$$r_{Z_2}^{-\kappa_2'} = \frac{(1-d_2)\bullet\bar{C}}{z_2\bullet\left[(1-d_2)\bullet\bar{C}\right]^{\varphi_2'}\bullet\left[(1-n_1-n_2)\bullet\bar{N}\right]^{\gamma_2'}} \qquad \textbf{(C-3B')}$$

$$r_{Z_2} = \left\{z_2\bullet\left[(1-d_2)\bullet\bar{C}\right]^{\varphi_2'-1}\bullet\left[(1-n_1-n_2)\bullet\bar{N}\right]^{\gamma_2'}\right\}^{\frac{1}{\kappa_2'}}$$

$$n_1\bar{N} = \left\{a_1\cdot(d_1\bar{C})^{\alpha_1}\bullet\left(P_{D_1}+r_{D_1}\right)^{-\varepsilon_1}\right\} - \left\{a_1'\bullet\left[(1-d_1)\bar{C}\right]^{\alpha_1'}\bullet\bar{N}^{\theta_1'}\bullet r_{Z_1}^{-\varepsilon_1'}\right\}$$

$$\left(P_{D_1}+r_{D_1}\right)^{-\varepsilon_1} = \frac{n_1\bar{N}+a_1'\bullet\left[(1-d_1)\bar{C}\right]^{\alpha_1'}\bullet\bar{N}^{\theta_1'}\bullet r_{Z_1}^{-\varepsilon_1'}}{a_1\bullet(d_1\bar{C})^{\alpha_1}}$$

$$P_{D_1} = \left\langle\frac{a_1\bullet(d_1\bar{C})^{\alpha_1}}{n_1\bar{N}+a_1'\bullet\left[(1-d_1)\bar{C}\right]^{\alpha_1'}\bullet\bar{N}^{\theta_1'}\bullet r_{Z_1}^{-\varepsilon_1'}}\right\rangle^{\frac{1}{\varepsilon_1}} - r_{D_1} \qquad \textbf{(C-1A')}$$

$$= \left\langle\frac{a_1\bullet(d_1\bar{C})^{\alpha_1}}{n_1\bar{N}+a_1'\bullet\left[(1-d_1)\bar{C}\right]^{\alpha_1'}\bullet\bar{N}^{\theta_1'}\bullet\left\{z_1\bullet\left[(1-d_1)\bullet\bar{C}\right]^{\varphi_1'-1}\bullet\left[(1-n_1)\bar{N}\right]^{\gamma_1'}\right\}^{-\frac{\varepsilon_1'}{\kappa_1'}}}\right\rangle^{\frac{1}{\varepsilon_1}}$$

$$-\left[b_1\bullet(d_1\bar{C})^{\varphi_1-1}\bullet(n_1\bar{N})^{\gamma_1}\right]^{\frac{1}{\kappa_1}}$$

$$n_2\bar{N}=\left\{a_2\bullet\left(d_2\bar{C}\right)^{\alpha_2}\bullet\left(n_1\bar{N}\right)^{\theta_2}\bullet\left(P_{D_2}+r_{D_2}\right)^{-\varepsilon_2}\right\}$$
$$-\left\{a_2'\bullet\left[\left(1-d_2\right)\bar{C}\right]^{\alpha_2'}\bullet\left[\left(1-n_1\right)\bar{N}\right]^{\theta_2'}\bullet r_{Z_2}^{-\varepsilon_2'}\right\}$$

$$\left(P_{D_2}+r_{D_2}\right)^{-\varepsilon_2}=\frac{n_2\bar{N}+a_2'\bullet\left[\left(1-d_2\right)\bar{C}\right]^{\alpha_2'}\bullet\left[\left(1-n_1\right)\bar{N}\right]^{\theta_2'}\bullet r_{Z_2}^{-\varepsilon_2'}}{a_2\bullet\left(d_2\bar{C}\right)^{\alpha_2}\bullet\left(n_1\bar{N}\right)^{\theta_2}}$$

$$P_{D_2}=\left\langle\frac{a_2\bullet\left(d_2\bar{C}\right)^{\alpha_2}\bullet\left(n_1\bar{N}\right)^{\theta_2}}{n_2\bar{N}+a_2'\bullet\left[\left(1-d_2\right)\bar{C}\right]^{\alpha_2'}\bullet\left[\left(1-n_1\right)\bar{N}\right]^{\theta_2'}\bullet r_{Z_2}^{-\varepsilon_2'}}\right\rangle^{\frac{1}{\varepsilon_2}}-r_{D_2} \tag{C-1B'}$$

$$=\left\langle\frac{a_2\bullet\left(d_2\bar{C}\right)^{\alpha_2}\bullet\left(n_1\bar{N}\right)^{\theta_2}}{n_2\bar{N}+a_2'\bullet\left[\left(1-d_2\right)\bar{C}\right]^{\alpha_2'}\bullet\left[\left(1-n_1\right)\bar{N}\right]^{\theta_2'}\bullet\left\{z_2\bullet\left[\left(1-d_2\right)\bullet\bar{C}\right]^{\varphi_2'-1}\bullet\left[\left(1-n_1-n_2\right)\bullet\bar{N}\right]^{\gamma_2'}\right\}^{-\frac{\varepsilon_2'}{\kappa_2'}}}\right\rangle^{\frac{1}{\varepsilon_2}}$$
$$-\left\{b_2\bullet\left(d_2\bar{C}\right)^{\varphi_2-1}\bullet\left[\left(n_1+n_2\right)\bar{N}\right]^{\gamma_2}\right\}^{\frac{1}{\kappa_2}}$$

$$\left(1-n_1-n_2\right)\bar{N}=\left\{a_3\bullet\bar{C}^{\alpha_3}\bullet\left(n_1+n_2\right)\bar{N}^{\theta_3}\bullet\left(P_{D_3}+r_{D_3}\right)^{-\varepsilon_3}\right\}$$

$$\left(P_{D_3}+r_{D_3}\right)^{-\varepsilon_3}=\frac{\left(1-n_1-n_2\right)\bar{N}}{a_3\bullet\bar{C}^{\alpha_3}\bullet\left(n_1+n_2\right)\bar{N}^{\theta_3}}$$

(C-1C′)

$$P_{D_3}=\left\{\frac{a_3\bullet\bar{C}^{\alpha_3}\bullet\left[\left(n_1+n_2\right)\bar{N}\right]^{\theta_3}}{\left(1-n_1-n_2\right)\bar{N}}\right\}^{\frac{1}{\varepsilon_3}}-r_{D_3}$$
$$=\left[\frac{a_3\left(n_1+n_2\right)^{\theta_3}}{\left(1-n_1-n_2\right)}\bullet\bar{C}^{\alpha_3}\bullet\bar{N}^{\theta_3-1}\right]^{\frac{1}{\varepsilon_3}}-\left(b_3\bullet\bar{C}^{\varphi_3-1}\bullet\bar{N}^{\gamma_3}\right)^{\frac{1}{\kappa_3}}$$

So, the hardware manufacturer's problem reduces to

Given n_1, n_2, and n_3, choose d_1, d_2, and d_3 to MAX

$$\begin{aligned}\Pi^{HW} = &\left\{\left[P_{D_1} - c\right] \bullet n_1\bar{N}\right\} + \left\{\delta \bullet \left[P_{D_2} - c \bullet (1-n_1)\right] \bullet n_2\bar{N}\right\} \\ &+ \left\{\delta^2 \bullet \left[P_{D_3} - c \bullet (1-n_1-n_2)\right] \bullet (1-n_1-n_2)\bar{N}\right\}\end{aligned} \quad \text{(HW-3')}$$

subject to the following:

$$\begin{aligned}P_{D_1} = &\left\langle \frac{a_1 \bullet (d_1\bar{C})^{\alpha_1}}{n_1\bar{N} + a_1' \bullet \left[(1-d_1)\bar{C}\right]^{\alpha_1'} \bullet \bar{N}^{\theta_1'} \bullet \left\{z_1 \bullet \left[(1-d_1) \bullet \bar{C}\right]^{\varphi_1'-1} \bullet \left[(1-n_1)\bar{N}\right]^{\gamma_1'}\right\}^{-\frac{\varepsilon_1'}{\kappa_1'}}} \right\rangle^{\frac{1}{\varepsilon_1}} \\ &- \left[b_1 \bullet (d_1\bar{C})^{\varphi_1-1} \bullet (n_1\bar{N})^{\gamma_1}\right]^{\frac{1}{\kappa_1}}\end{aligned} \quad \text{(C-1A')}$$

$$\begin{aligned}P_{D_2} = &\left\langle \frac{a_2 \bullet (d_2\bar{C})^{\alpha_2} \bullet (n_1\bar{N})^{\theta_2}}{n_2\bar{N} + a_2' \bullet \left[(1-d_2)\bar{C}\right]^{\alpha_2'} \bullet \left[(1-n_1)\bar{N}\right]^{\theta_2'} \bullet \left\{z_2 \bullet \left[(1-d_2) \bullet \bar{C}\right]^{\varphi_2'-1} \bullet \left[(1-n_1-n_2) \bullet \bar{N}\right]^{\gamma_2'}\right\}^{-\frac{\varepsilon_2'}{\kappa_2'}}} \right\rangle^{\frac{1}{\varepsilon_2}} \\ &- \left\{b_2 \bullet (d_2\bar{C})^{\varphi_2-1} \bullet \left[(n_1+n_2)\bar{N}\right]^{\gamma_2}\right\}^{\frac{1}{\kappa_2}}\end{aligned} \quad \text{(C-1B')}$$

$$P_{D_3} = \left[\frac{a_3(n_1+n_2)^{\theta_3}}{(1-n_1-n_2)} \bullet \bar{C}^{\alpha_3} \bullet \bar{N}^{\theta_3-1}\right]^{\frac{1}{\varepsilon_3}} - \left(b_3 \bullet \bar{C}^{\varphi_3-1} \bullet \bar{N}^{\gamma_3}\right)^{\frac{1}{\kappa_3}} \quad \text{(C-1C')}$$

and subject to the restrictions on transition variables, (T-1A) and (T-2A).

A.5.2 Derivation of the Content Provider's Problem

The maximization problem for the content provider in the three-period model is:

Given d_1, d_2, and d_3, choose n_1, n_2, and n_3 to MAX

$$\begin{aligned}\Pi^{SW} = & \left(r_{D_1} - c_{D_1}\right)\bullet d_1\bar{C} - F_D + \left(r_{Z_1} - c_{Z_1}\right)\bullet\left(1-d_1\right)\bullet\bar{C} \\ & +\delta\bullet\left[\left(r_{D_2} - c_{D_2}\right)\bullet d_2\bar{C} + \left(r_{Z_2} - c_{Z_2}\right)\bullet\left(1-d_2\right)\bullet\bar{C}\right] \\ & +\delta^2\bullet\left[\left(r_{D_3} - c_{D_3}\right)\bullet\bar{C}\right]\end{aligned} \quad \text{(SW-3)}$$

subject to the following:

First-, second-, and third-period user demand functions for hardware:

$$\begin{aligned}n_1\bar{N} = & \left\{a_1\bullet\left(d_1\bar{C}\right)^{\alpha_1}\bullet\left(P_{D_1} + r_{D_1}\right)^{-\varepsilon_1}\right\} \\ & -\left\{a_1'\bullet\left[\left(1-d_1\right)\bar{C}\right]^{\alpha_1'}\bullet\bar{N}^{\theta_1'}\bullet r_{Z_1}^{-\varepsilon_1'}\right\}\end{aligned} \quad \text{(C-1A)}$$

$$\begin{aligned}n_2\bar{N} = & \left\{a_2\bullet\left(d_2\bar{C}\right)^{\alpha_2}\bullet\left(n_1\bar{N}\right)^{\theta_2}\bullet\left(P_{D_2} + r_{D_2}\right)^{-\varepsilon_2}\right\} \\ & -\left\{a_2'\bullet\left[\left(1-d_2\right)\bar{C}\right]^{\alpha_2'}\bullet\left[\left(1-n_1\right)\bar{N}\right]^{\theta_2'}\bullet r_{Z_2}^{-\varepsilon_2'}\right\}\end{aligned} \quad \text{(C-1B)}$$

$$\left(1-n_1-n_2\right)\bar{N} = \left\{a_3\bullet\bar{C}^{\alpha_3}\bullet\left[\left(n_1+n_2\right)\bar{N}\right]^{\theta_3}\bullet\left(P_{D_3} + r_{D_3}\right)^{-\varepsilon_3}\right\} \quad \text{(C-1C)}$$

First-, second-, and third-period user demand functions for new system software:

$$D_1 = d_1\bar{C} = b_1\bullet\left(d_1\bar{C}\right)^{\varphi_1}\bullet\left(n_1\bar{N}\right)^{\gamma_1}\bullet r_{D_1}^{-\kappa_1} \quad \text{(C-2A)}$$

$$D_2 = d_2\bar{C} = b_2\bullet\left(d_2\bar{C}\right)^{\varphi_2}\bullet\left[\left(n_1+n_2\right)\bar{N}\right]^{\gamma_2}\bullet r_{D_2}^{-\kappa_2} \quad \text{(C-2B)}$$

$$D_3 = d_3\bar{C} = b_3\bullet\left(d_3\bar{C}\right)^{\varphi_3}\bullet\bar{N}^{\gamma_3}\bullet r_{D_3}^{-\kappa_3} \quad \text{(C-2C)}$$

First- and second-period user demand functions for old system software:

$$Z_1 = (1-d_1)\bullet\bar{C} = z_1\bullet\left[(1-d_1)\bullet\bar{C}\right]^{\varphi_1'}\bullet\left[(1-n_1)\bar{N}\right]^{\gamma_1'}\bullet r_{Z_1}^{-\kappa_1'} \qquad \textbf{(C-3A)}$$

$$Z_2 = (1-d_2)\bullet\bar{C} = z_2\bullet\left[(1-d_2)\bullet\bar{C}\right]^{\varphi_2'}\bullet\left[(1-n_1-n_2)\bullet\bar{N}\right]^{\gamma_2'}\bullet r_{Z_2}^{-\kappa_2'} \qquad \textbf{(C-3B)}$$

Restrictions on transition variables:

$$n_1 + n_2 + n_3 = 1 \qquad \textbf{(T-1A)}$$

$$0 \le d_1 \le d_2 \le d_3 = 1 \qquad \textbf{(T-2A)}$$

So the software provider's problem reduces to

Given d_1, d_2, and d_3, choose n_1, n_2, and n_3 to MAX

$$\begin{aligned}\Pi^{SW} &= \left(r_{D_1} - c_{D_1}\right)\bullet d_1\bar{C} - F_D + \left(r_{Z_1} - c_{Z_1}\right)\bullet(1-d_1)\bar{C} \\ &+\delta\bullet\left[\left(r_{D_2} - c_{D_2}\right)\bullet d_2\bar{C} + \left(r_{Z_2} - c_{Z_2}\right)\bullet(1-d_2)\bar{C}\right] \\ &+\delta^2\bullet\left[\left(r_{D_3} - c_{D_3}\right)\bullet\bar{C}\right]\end{aligned}$$

$$\begin{aligned}&= \left\{\left[b_1\bullet(d_1\bar{C})^{\varphi_1-1}\bullet(n_1\bar{N})^{\gamma_1}\right]^{\frac{1}{\kappa_1}} - c_{D_1}\right\}\bullet d_1\bar{C} - F_D + \left(\left\{z_1\bullet\left[(1-d_1)\bullet\bar{C}\right]^{\varphi_1'-1}\bullet\left[(1-n_1)\bar{N}\right]^{\gamma_1'}\right\}^{\frac{1}{\kappa_1'}} - c_{Z_1}\right)\bullet(1-d_1)\bar{C} \\ &+\delta\bullet\left\langle\left(\left\{b_2\bullet(d_2\bar{C})^{\varphi_2-1}\bullet\left[(n_1+n_2)\bar{N}\right]^{\gamma_2}\right\}^{\frac{1}{\kappa_2}} - c_{D_2}\right)\bullet d_2\bar{C} + \left(\left\{z_2\bullet\left[(1-d_2)\bullet\bar{C}\right]^{\varphi_2'-1}\bullet\left[(1-n_1-n_2)\bullet\bar{N}\right]^{\gamma_2'}\right\}^{\frac{1}{\kappa_2'}} - c_{Z_2}\right)\bullet(1-d_2)\bar{C}\right\rangle \\ &+\delta^2\bullet\left\langle\left[\left(b_3\bullet\bar{C}^{\varphi_3-1}\bullet\bar{N}^{\gamma_3}\right)^{\frac{1}{\kappa_3}} - c_{D_3}\right]\bullet\bar{C}\right\rangle\end{aligned} \qquad \textbf{(SW-3')}$$

subject to the restrictions on transition variables, (T-1A) and (T-2A).

A.5.3 Equilibrium

The equilibrium of the two-period model is a set of prices for the hardware manufacturer and content provider, for which the following two conditions simultaneously hold:

- When the content provider chooses its equilibrium set of content allocations, $n^* = \{n_1^*, n_2^*, \ldots\}$, then the hardware manufacturer earns maximum profits by choosing its equilibrium set of portions of user adoption, $d^*=\{d_1^*, d_2^*, \ldots\}$; that is, if the content provider makes its equilibrium set of content allocations, n^*, then there is no other set of portions of user adoption, d', that will earn the hardware manufacturer greater profits than those earned by choosing its equilibrium portions of user adoption, d^*:

$$\Pi^{HW}(d^*, n^*) \geq \Pi^{HW}(d', n^*) \text{ for all } d'$$

- When the hardware manufacturer chooses its equilibrium set of portions of user adoption, d^*, the content provider earns maximum profits by choosing its equilibrium set of allocations, n^*; that is, if the hardware manufacturer chooses portions of user adoption, d^*, then there is no other set of content allocations, n', that will earn the content provider greater profits than those earned by choosing its equilibrium set of content allocations, n^*:

$$\Pi^{SW}(d^*, n^*) \geq \Pi^{SW}(d^*, n') \text{ for all } n'$$

APPENDIX B

Further Information

The following resources provide further information on particular issues relating to technology adoption.

B.1 Adoption of VHS versus Betamax

Belson, Ken. 2006. In Sony's Stumble, the Ghost of Betamax. *New York Times*, February 26.

Cusumano, Michael A., Yiorgos Mylonadis, and Richard S. Rosenbloom. 1992. Strategic Maneuvering and Mass-Market Dynamics: The Triumph of VHS over Beta. *Business History Review* 66, no. 1 (Spring): 51–94.

Digital Silence. 2001. Recordable DVD's—Betamax/VHS All Over Again. February 26. www.d-silence.com/feature.php?id=222.

Fasoldt, Al. 1988. How Sony Killed Betamax. *Syracuse Newspapers*. http://aroundcny.com/technofile/texts/howbetadied.htm.

Media College. The Betamax vs VHS Format War. www.mediacollege.com/video/format/compare/betamax-vhs.html.

Moulding, Heldge. 1996. The Decline and Fall of Betamax. AFU White Paper. http://tafkac.org/products/beta_vs_vhs.html.

Olleros, Xavier. 2003. Betamax Revisited: A Contextual View of the Battle for the Home VCR Market. Université du Québec à Montréal. www.sba.muohio.edu/abas/2003/brussels/olleros_betamax%20revisited.pdf.

Pilieci, Vito. 2005. VHS Format Heading to Extinction. *CanWest News Service*, October 20. www.canada.com/technology/story.html?id=89497774-3f47-46fa-9aef-3ab3d392bdd9&page=1.

B.2 Adoption of Next-Generation DVD

Belson, Ken. 2005. DVD Fight Intensifies: Microsoft and Intel to Back Toshiba Format. *New York Times,* September 27.

———. 2006. As DVD Sales Slow, Hollywood Hunts for a New Cash Cow. *New York Times,* June 13.

Edwards, Cliff. 2006. Next-Gen DVD's Blurry Picture. *BusinessWeek,* March 27.

Grover, Ronald, and Cliff Edwards. 2007. Next-Gen DVDs: Advantage, Sony. *BusinessWeek,* December 6.

Guth, Robert A., and Don Clark. 2005. Microsoft, Intel Back HD-DVD As Next Format for Data Storage. *Wall Street Journal,* September 27.

McGrath, Dylan. 2006. "Teardown" Finds Toshiba Taking a Loss on HD DVD Player. *EETimes,* June 23.

Pogue, David. 2006. Why the World Doesn't Need Hi-Def DVD's. *New York Times,* May 11.

Schuman, Joseph. 2005. Paramount's Shift Mixes Up the DVD Format Wars. *Wall Street Journal,* October 3.

Wingfield, Nick. 2006. Format Faceoff: Bringing the DVD War Home: Blu-ray Player and Titles Hit Stores, Taking On HD-DVD; Is It Time to Choose Sides? *Wall Street Journal,* June 20.

B.3 Adoption of HDTV

Bayus, Barry L. 1993. High-Definition Television: Assessing Demand Forecasts for a Next Generation Consumer Durable. *Management Science* 39, no. 11 (November): 1319–33.

Benjamin, David. 2005. Falling System Prices Boost HDTV Prospects. *EETimes,* September 19. www.eetimes.com/showArticle.jhtml?articleID=170703864.

Darlin, Damon. 2005. Big TVs at Smaller Prices: Steep Price Drops Expected for Fall Sale Season. *New York Times,* August 20.

Davis, Bob. 1989. Will High-Definition TV Be a Turn-Off? Perhaps, Thanks to Cost of Sets and Screen Size. *Wall Street Journal,* January 20.

Freeman, Dr. Jonathan, and Dr. Jane Lessiter. 2002. Using Attitude Based Segmentation to Better Understand Viewers' Usability Issues with Digital and Interactive TV. University of London, New Cross, London, UK. www.brighton.ac.uk/interactive/euroitv/euroitv03/Papers/Paper3.pdf.

Gupta, Sachin, Dipak C. Jain, and Mohanbir S. Sawhney. 1999. Modeling the Evolution of Markets with Indirect Network Externalities: Application to Digital Television. Special Issue on Managerial Decision Making. *Marketing Science* 18, no. 3: 396–416.

Jacobs, Joanne. 2001. DTV2001: Services, Adoption, Investment and Policy Issues Arising from the First Year of Digital Television in Australia. Communication Research Forum 2001. www.archive.dcita.gov.au/2004/10/issue_4/communications_research_forum.

Ramstad, Evan, and Phred Dvorak. 2005. Wave of Price Cutting Hits Flat-Panel TV Sets. *Wall Street Journal*, December 29.

Rast, Robert M. The Dawn of Digital TV. *IEEE Spectrum Online*. www.spectrum.ieee.org/print/1911.

RedHerring.com. 2005. High-Def Video Catches On. July 20. www.redherring.com/Home/12838.

Schatz, Amy. 2005. Crossing the Digital Television Divide. *Wall Street Journal*, July 6.

Steinberg, Brian. 2005. TV On-Demand May Make Ads More Targeted. *Wall Street Journal*, November 9.

Taub, Eric A. 2006. Move Over, HD-TV. Now There's HD Radio, Too. *New York Times*, January 23.

van Grinsven, Lucas, and Georgina Prodhan. 2005. High Definition TV Starts Slowly, Makers Hopeful. *Red Orbit*, September 1. www.redorbit.com/news/business/227024/high_definition_tv_starts_slowly_makers_hopeful/index.html.

Wingfield, Nick. 2006. A New Reason to Buy a High-Definition TV. *Wall Street Journal*, January 4.

Yang, Catherine, Heather Green, and Tom Lowry. 2005. Everybody Wants a Piece of the Air. *Business Week*, July 4. www.businessweek.com/magazine/content/05_27/b3941040_mz011.htm.

B.4 Adoption of Consumer Durables

Bass, Frank M. 1969. A New Product Growth for Model Consumer Durables. *Management Science* 15, no. 5 (January): 215–27.

———. 1980. The Relationship between Diffusion Rates, Experience Curves, and Demand Elasticities for Consumer Durable Technological Innovations, Part 2: Interfaces Between Marketing and Economics. *Journal of Business* 53, no. 3 (July): S51–S67.

Bayus, Barry L. 1992. The Dynamic Pricing of Next Generation Consumer Durables. *Marketing Science* 11, no. 3 (Summer): 251–65.

Coplan, Judson. 2006. Diagnosing the DVD Disappointment: A Life Cycle View. *Stern School of Business.*

Golder, Peter N., and Gerard J. Tellis. 1997. Will It Ever Fly? Modeling the Takeoff of Really New Consumer Durables. *Marketing Science* 16, no. 3: 256–70.

Parker, Philip M., and Ramya Neelamegham. 1997. Price Elasticity Dynamics over the Product Life Cycle: A Study of Consumer Durables. *Marketing Letters* 8, no. 2: 205–16.

Russell, Thomas. 1980. Comments on "The Relationship Between Diffusion Rates, Experience Curves, and Demand Elasticities for Consumer Durable Technological Innovations," Part 2: Interfaces Between Marketing and Economics. *Journal of Business* 53, no. 3 (July): S69–S73.

B.5 Networks and Network Effects

Briscoe, Bob, Andrew Odlyzko, and Benjamin Tilly. 2006. Metcalfe's Law Is Wrong. *IEEE Spectrum Online*, July. http://spectrum.ieee.org/jul06/4109.

Carlton, Dennis W., and J. Mark Klamer. 1983. The Need for Coordination among Firms, with Special Reference to Network Industries. *University of Chicago Law Review* 50 (Spring): 446–65.

Church, Jeffrey, and Neil Gandal. 1992. Network Effects, Software Provision and Standardization. *Journal of Industrial Economics* XL (March): 85–103.

Church, Jeffrey, Neil Gandal, and David Krause. 2003. Indirect Network Effects and Adoption Externalities. CEPR Discussion Papers 3738, December 19. http://ideas.repec.org/r/bla/jindec/v40y1992i1p85-103.html.

Cueller, Steven S. 2002. The New Economy, Network Effects and Market Structure. Paper presented at the National Business and Economic Society Conference, Maui, Hawaii, March 8.

Economides, Nicholas. 1993. Network Economics with Application to Finance. *Financial Markets, Institutions & Instruments* 2, no. 5: 89–97.

———. 1996. The Economics of Networks. *International Journal of Industrial Organization* 14, no. 6: 673–99.

Economides, Nicholas, and Steven C. Salop. 1992. Competition and Integration among Complements, and Network Market Structure. *Journal of Industrial Economics* 40, no. 1 (March): 105–23.

Economides, Nicholas, and V. Brian Viard. 2005. Pricing of Complementary Goods and Network Effects. Working Paper #05-04, Stern School of Business, New York University.

Goldenberg, Jacob, Barak Libai, and Eitan Muller. 2002. Is the Bandwagon Rolling? The Chilling Effect of Network Externalities on New Product Growth. November. http://recanati.tau.ac.il/Eng/Index.asp?ArticleID=334&CategoryID=274&Page=1.

Greenstein, Shane. 1994. Invisible Hand versus Invisible Advisors: Coordination Mechanisms in Economic Networks. University of Illinois, Urbana/Champaign, Champaign, IL. www.columbia.edu/dlc/wp/citi/citi489.html.

Katz, Michael L., and Carl Shapiro. 1986. Technology Adoption in the Presence of Network Externalities. *Journal of Political Economy* 94, no. 41: 822–41.

———. 1994. Systems Competition and Network Effects. *Journal of Economic Perspectives* 8, no. 2: 93–115.

Liebowitz, S. J., and Stephen E. Margolis. 1994. Network Externality: An Uncommon Tragedy. *Journal of Economic Perspectives* 8, no. 2: 133–50.

———. 1995. Are Network Externalities a New Source of Market Failure? *Research in Law and Economics* 17:1–22.

Odlyzko, Andrew, and Benjamin Tilly. 2005. A Refutation of Metcalfe's Law and a Better Estimate for the Value of Networks and Network Interconnections. March 2. www.dtc.umn.edu/~odlyzko/doc/metcalfe.pdf.

Ross, Philip E. 2003. 5 Commandments. *IEEE Spectrum Online*. www.spectrum.ieee.org/dec03/3752.

Shankar, Venkatesh, and Barry L. Bayus. 2002. Network Effects and Competition: An Empirical Analysis of the Home Video Game Industry. University of Maryland, College Park, MD. http://papers.ssrn.com/sol3/papers.cfm?abstract_id=296534.

Shy, Oz. 2001. *The Economics of Network Industries*. Cambridge: Cambridge University Press.

Srinivasan, Raji , Gary L. Lilien, and Arvind Rangaswamy. 2001. When Being First Is Not Enough: Network Externalities and the Survival of Pioneers. ISBM Report 4-2002, December, Pennsylvania State University.

Stremersch, Stefan, Gerard J. Tellis, Philip Hans Franses, and Jeroen L. G.Binken. 2007. Indirect Network Effects in New Product Growth. *Journal of Marketing* 71, no. 3 (July): 52–74. www.atypon-link.com/AMA/doi/abs/10.1509/jmkg.71.3.52?journalCode=jmkg.

Varian, Hal. 2001. High Technology Industries and Market Structure, with Commentaries by Erik Brynjolfsson and Kevin M. Murphy. http://people.ischool.berkeley.edu/~hal/Papers/structure/.

B.6 Lock-in and Path Dependence

Antonelli, Cristiano. 2000. Path Dependence, Localized Technological Change and the Quest for Dynamic Efficiency, University of Torino, Italy. www.fondazionerosselli.it/DocumentFolder/Path_dependence_change.doc.

Arthur, W. Brian. 1989. Competing Technologies, Increasing Returns, and Lock-in by Historical Events. *Economic Journal* 99, no. 394 (March): 116–31.

———. 1996. Increasing Returns and the New World of Business. *Harvard Business Review*, July–August.

David, Paul A. 2000. Path Dependence, Its Critics, and the Quest for "Historical Economics." All Souls College, Oxford, and Stanford University. www.econ.stanford.edu/faculty/workp/swp00011.pdf.

Liebowitz, S. J., and Stephen E. Margolis. 1995. Path Dependence, Lock-In, and History. *Journal of Law, Economics and Organization* 11, no. 1: 205–26.

B.7 Standardization and Compatibility

Economides, Nicholas. 1991. Compatibility and Market Structure. Discussion Paper EC-91-16, Stern School of Business, New York University.

Farrell, Joseph, and Garth Saloner 1985. Standardization, Compatibility, and Innovation. *RAND Journal of Economics* 16, no. 1 (Spring): 70–83.

———. 1986. Installed Base and Compatibility: Innovation, Product Preannouncements, and Predation. *American Economic Review* 76, no. 5 (December): 940–55.

Gandal, Neil. 2002. Compatibility, Standardization, and Network Effects: Some Policy Implications. *Oxford Review of Economic Policy* 18, no. 1 (Spring): 80–91.

Katz, Michael L., and Carl Shapiro. 1985. Network Externalities, Competition, and Compatibility. *American Economic Review* 75, no. 3: 424–40.

Matutes, Carmen, and Pierre Regibeau. 1988. "Mix and Match": Product Compatibility without Network Externalities. *RAND Journal of Economics* 19, no. 2: 221–34.

Nelson, Matthew L., and Michael J. Shaw. 2001. Adoption of Technology Standards in Supply Chains: A Case of Adopting Rosettanet Standards. Paper submitted to DIGIT—2001 IT Adoption & Diffusion Workshop (December 16). http://citebm.business.uiuc.edu/IT_cases/rosettanet.pdf.

Stango, Victor. 2004. The Economics of Standards Wars. *Review of Network Economics* 3, no. 1 (March): 1–19.

B.8 Innovation and Adoption of New Technologies

Agarwal, Rajshree, and Barry L. Bayus, 2002. The Market Evolution and Sales Takeoff of Product Innovations. *Management Science* 48, no. 8 (August): 1024–41.

Chakravorti, Bhaskar. 2003. *The Slow Pace of Fast Change.* Boston: Harvard Business School Press.

———. 2004. The Role of Adoption Networks in the Success of Innovations. *Technology in Society* 26 (April–August): 469–82.

Chandy, Rajesh K., and Gerard J. Tellis. 2000. The Incumbent's Curse? Incumbency, Size, and Radical Product Innovation. *Journal of Marketing* 64, no. 3 (July): 1–17.

ConStat, Inc. 2001. ConStat's Technology Adoption Model: The Framework for the WISDM Program. www.constat.com/wisdm/ConStat_Tech_Adoption_Model.pdf. (Note: The link no longer seems to be active; the company's website is www.constat .com/index.php.)

Danaher, Peter J., Bruce G. S. Hardie, and William P. Putsis Jr. 2001. Marketing-Mix Variables and the Diffusion of Successive Generations of a Technological Innovation. *Journal of Marketing Research* 38, no. 4 (November): 501–14.

Enderle, Rob. 2008. Analyst Angle: How to Beat the iPhone and the Blackberry. *RCR Wireless News*, March 24.

Fichman, Robert G. 1992. Information Technology Diffusion: A Review of Empirical Research. MIT Sloan School of Management, Cambridge, MA. Paper presented at the Proceedings of the Thirteenth International Conference on Information Systems. www2.bc.edu/~fichman/Fichman_1992_ICIS_IT_Diff_Review.pdf.

———. 1999. The Diffusion and Assimilation of Information Technology Innovations, in *Framing the Domains of IT Management: Projecting the Future through the Past*, ed. R. W. Zmud. Cincinnati, OH: Pinnaflex Educational Resources, Inc.

Gort, Michael, and Steven Klepper. 1982. Time Paths in the Diffusion of Product Innovations. *Economic Journal* 92, no. 367 (September): 630–53.

Knowledge@Wharton. 2006. "Influentials" and "Imitators": How to Better Forecast the Sale of New Products, September 6.

Lin, Zhangxi, and Nirup M. Menon. 2003. A Transaction Cost Nash Equilibrium Model for the Adoption of Innovative Information Technology. Texas Tech University, Lubbock, TX. zlin.ba.ttu.edu/papers/working/IT-Adopt7a.pdf.

Montaguti, Elisa, Sabine Kuester, and Thomas S. Robertson. 2002. Entry Strategy for Radical Product Innovations: A Conceptual Model and Propositional Inventory. *International Journal of Research in Marketing* 19:21–42.

Parker, Philip M. 1992. Price Elasticity Dynamics over the Adoption Life Cycle. *Journal of Marketing Research* 29, no. 3 (August): 358–67.

Pavlova, Anna. 1999. Adjustment Costs, Learning-by-Doing, and Technology Adoption under Uncertainty. CARESS Working Paper 99-07. www.econ.upenn.edu/Centers/ CARESS/.

Rogers, Everett M. 2003. *Diffusion of Innovations, Fifth Edition*. New York: Free Press.

van den Bulte, Christophe. 2000. New Product Diffusion Acceleration: Measurement and Analysis. *Marketing Science* 19, no. 4 (Autumn): 366–80.

Weiss, Allen M. 1994. The Effects of Expectations on Technology Adoption: Some Empirical Evidence. *Journal of Industrial Economics* 42, no. 4 (December): 341–60.

B.9 Product Lifecycles

Bayus, Barry L. 1998. An Analysis of Product Lifetimes in a Technologically Dynamic Industry. *Management Science* 44, no. 6 (June): 763–75.

Chandy, Rajesh K., and Gerard J. Tellis. 1998. Organizing for Radical Product Innovation: The Overlooked Role of Willingness to Cannibalize. *Journal of Marketing Research* 35, no. 4 (November): 474–87.

Gandal, Neil, Michael Kende, and Rafael Rob. 2000. The Dynamics of Technological Adoption in Hardware/Software Systems: The Case of Compact Disc Players. *RAND Journal of Economics* 31, no. 1: 43–61.

Huisman, Kuno J. M., and Peter M. Kort. 2004. Strategic Technology Adoption Taking into Account Future Technological Improvements: A Real Options Approach. *European Journal of Operational Research* 159, no. 3 (December): 705–28. www.sciencedirect.com/science?_ob=ArticleURL&_udi=B6VCT-49KSRYC-F&_user=10&_rdoc=1&_fmt=&_orig=search&_sort=d&view=c&_acct=C000050221&_version=1&_urlVersion=0&_userid=10&md5=009f523df00cb5d6029f0e6c5d4d37e3.

Norton, John A., and Frank M. Bass. 1987. A Diffusion Theory Model of Adoption and Substitution for Successive Generations of High-Technology Products. *Management Science* 33, no. 9 (September): 1069–86.

B.10 Critical Mass

Caillaud, Bernard, and Bruno Jullien. 2001. Chicken & Egg: Competing Matchmakers. April. Centre for Economic Policy Research (CEPR) Working Paper.

Coye, Molly Joel, MD, and Wade M. Aubrey, MD. 2003. The "Tipping Point" and Health Care Innovations: Advancing the Adoption of Beneficial Technologies. National

Committee for Health Care Conference, Washington, DC, January 27–28, 2003. www.nihcm.org/~nihcmor/pdf/Coye.pdf.

Economides, Nicholas, and Charles Himmelberg. 1995. Critical Mass and Network Size with Application to the US Fax Market. Discussion Paper EC-95-11, Stern School of Business, New York University.

Economist. 2005. Matchmakers and Trustbusters, December 8.

Evans, David S. 2003. Some Empirical Aspects of Multi-Sided Platform Industries. *Review of Network Economics* 2, no. 3 (September): 191–209.

Gladwell, Malcolm. 2002. *The Tipping Point.* Boston: Back Bay Books.

Lim, Byeong-Lak, Munkee Choi, and Myeong-Cheol Park. 2003. The Late Take-off Phenomenon in the Diffusion of Telecommunication Services: Network Effect and the Critical Mass. *Information Economics and Policy* 15: 537–57.

Moore, Geoffrey A. 1999. *Crossing the Chasm.* New York: Harper Collins.

Park, Sangin. 2004. Strategic Maneuvering and Standardization: Critical Advantage or Critical Mass? Seoul National University. http://econpapers.repec.org/paper/ecmfeam04/596.htm.

Rochet, Jean-Charles, and Jean Tirole. 2005. Two-Sided Markets: A Progress Report. November. http://ideas.repec.org/p/ide/wpaper/2145.html.

Witt, Ulrich. 1997. "Lock-in" v. "Critical Masses"—Industrial Change under Network Externalities. *International Journal of Industrial Organization* 15: 753–73.

B.11 First-Mover Advantages in Adoption of New Technologies

Golder, Peter N., and Gerard J. Tellis. 1993. Pioneer Advantage: Marketing Logic or Marketing Legend? *Journal of Marketing Research* 30, no. 2 (May): 158–70.

Kerin, Roger A., P. Rajan Varadarajan, and Robert A. Peterson. 1992. First Mover Advantage: A Synthesis, Conceptual Framework, and Research Propositions. *Journal of Marketing* 56, no. 4 (October): 33–52.

Shankar, Venkatesh, Gregory S. Carpenter, and Lakshman Krishnamurthi. 1998. Late Mover Advantage: How Innovative Late Entrants Outsell Pioneers. *Journal of Marketing Research* 35, no. 1 (February): 54–70.

Urban, Glen L., Theresa Carter, Steven Gaskin, and Zofia Mucha. 1986. Market Share Rewards to Pioneering Brands: An Empirical Analysis and Strategic Implications. *Management Science* 32, no. 6 (June): 645–59.

VanderWerf, Pieter A., and John F. Mahon. 1997. Meta-Analysis of the Impact of Research Methods on Findings of First-Mover Advantage. *Management Science* 43, no. 11 (November): 1510–19.

B.12 Social Networks and Technology Adoption

Goldenberg, Jacob, Barak Libai, and Eitan Muller. 2001. Talk of the Network: A Complex Systems Look at the Underlying Process of Word-of-Mouth. *Marketing Letters* 12, no. 3: 211–23.

Granovetter, Mark. 1983. The Strength of Weak Ties: A Network Theory Revisited. *Sociological Theory* 1: 201–33.

Konana, Prabhudev, and Sridhar Balasubramanian. 2005. The Socio-Economic-Psychological (SEP) Model of Technology Adoption and Usage: An Application to Online Investing. University of Texas at Austin, Austin, TX. *Decision Support Systems* 39, no. 3: 505–24 (2005).

Kraut, Robert E., Ronald E. Rice, Colleen Cool, and Robert S. Fish. 1998. Varieties of Social Influence: The Role of Utility and Norms in the Success of a New Communications Medium. *Organizational Science* 9, no. 4: 437–53.

Liebenstein, H. 1950. Bandwagon, Snob, and Veblen Effects in the Theory of Consumers' Demand. *Quarterly Journal of Economics* 64, no. 2: 183–207.

Ravid, Gilad, and Sheizaf Rafaeli. 2004. Asynchronous Discussion Groups as Small World and Scale Free Networks. *First Monday* 9, no. 9 (September). http://ebipol.p.lodz.pl/Content/1081/issues/issue9_9/ravid/index.html.

Rolfe, Meredith. 2004. Social Networks and Threshold Models of Collective Behavior. University of Chicago. December 10. http://home.uchicago.edu/~mrrolfe/net_threshold.v2.pdf.

Valente, Thomas W. 1996. Social Network Thresholds in the Diffusion of Innovations. *Social Networks* 18: 69–89.

Valente, Thomas W., and Robert K. Foreman. 1998. Integration and Radiality: Measuring the Extent of an Individual's Connectedness and Reachability in a Network. *Social Networks* 20: 89–105.

Vigneron, Franck, and Lester W. Johnson. 1999. A Review and a Conceptual Framework of Prestige-Seeking Consumer Behavior. *Academy of Marketing Science Review* 1999, no. 1.

References

Agarwal, Rajshree, and Barry L. Bayus, 2002. The Market Evolution and Sales Takeoff of Product Innovations. *Management Science* 48, no. 8 (August): 1024–41.

Apple Computer, Inc. 2001. Apple Introduces iTunes — World's Best and Easiest to Use Jukebox Software. Press release, January 9, 2001. www.apple.com/pr/library/2001/jan/09itunes.html.

Aumann, Robert J. 1987. Game Theory. *The New Palgrave Dictionary of Economics*, vol. 2, ed. Steven N. Durlauf and Lawrence E. Blume. New York: Palgrave Macmillan, 460–82.

Bass, Frank M. 1969. A New Product Growth for Model Consumer Durables. *Management Science* 15, no. 5 (January): 215–27.

Bayus, Barry L. 1992. The Dynamic Pricing of Next Generation Consumer Durables. *Marketing Science* 11, no. 3 (Summer): 251–65.

———. 1998. An Analysis of Product Lifetimes in a Technologically Dynamic Industry. *Management Science* 44, no. 6 (June): 763–75.

Ben-David, S., A. Borodin, R. Karp, G. Tardos, and A. Wigderson. 1994. On the Power of Randomization in On-Line Algorithms. *Algorithmica* 11, no. 1 (January): 2–14.

Brenn, Max. 2008. Apple's AppStore to Open on Friday with 500 Applications. *eFlux Media*, July 10. www.efluxmedia.com/news_Apples_AppStore_To_Open_On_Friday_With_500_Applications_20140.html.

Camerer, Colin F. 2003. *Behavioral Game Theory: Experiments in Strategic Interaction*. Princeton, NJ: Princeton University Press.

Cole, Robert. 2007. Move to Digital Movie Projection Is Going Slowly. *Kansas City Star,* December 3.

Coplan, Judson. 2006. Diagnosing the DVD Disappointment: A Life Cycle View. *Stern School of Business.*

Danaher, Peter J., Bruce G. S. Hardie, and William P. Putsis Jr. 2001. Marketing-Mix Variables and the Diffusion of Successive Generations of a Technological Innovation. *Journal of Marketing Research* 38, no. 4 (November): 501–14.

Dealerscope. 2007. DVD Sales Drop for the First Time, December 10. www.dealerscope.com/article/83000-83999/83290_1.html.

Dillon, Pat. 1998. The Next Small Thing. *Fast Company*, no. 15 (May): 97.

Downs, Anthony. 1957. *An Economic Theory of Democracy.* New York: Harper.

Eran, Daniel. 2006. Newton Lessons for Apple's New Platform, December 11. www.roughlydrafted.com/RD/Q4.06/600D65E6-A31E-45CA-AFC5-42BC253F5337.html.

Fraker, S. 1984. High Speed Management for the High Tech Age. *Fortune* 62 (March 5).

Holson, Laura M. 2008. Hoping to Make Phone Buyers Flip. *New York Times*, February 29.

Hormby, Tom. 2007. Early History of Palm (Part I). *Silicon User*, October 21.

McBride, Sarah. 2005. Hollywood's Digital Delay: Studios Finally Agree to Pay to Computerize Projectors but Theaters Have Qualms. *Wall Street Journal*, October 27.

Maynard Smith, John. 1982. *Evolution and the Theory of Games.* Cambridge: Cambridge University Press.

Moore, Geoffrey A. 1999. *Crossing the Chasm*. New York: Harper Collins Publishers, Inc.

Norton, John A., and Frank M. Bass. 1987. A Diffusion Theory Model of Adoption and Substitution for Successive Generations of High-Technology Products. *Management Science* 33, no. 9 (September): 1069–86.

Odlyzko, Andrew, and Benjamin Tilly. 2005. A Refutation of Metcalfe's Law and a Better Estimate for the Value of Networks and Network Interconnections. March 2.

http://74.125.95.132/search?q=cache:btSiCeZTZ9wJ:www.dtc.umn.edu/~odlyzko/doc/metcalfe.pdf+A+refutation+of+Metcalfe%E2%80%99s+Law+and+a+better+estimate+for+the+value+of+networks+and+network+interconnections&hl=en&ct=clnk&cd=1&gl=us.

Parker, Philip M. 1992. Price Elasticity Dynamics over the Adoption Life Cycle. *Journal of Marketing Research* 29, no. 3 (August): 358–67.

Parker, Philip M., and Ramya Neelamegham. 1997. Price Elasticity Dynamics over the Product Life Cycle: A Study of Consumer Durables. *Marketing Letters* 8, no. 2: 205–16.

PRWeb. 2008. Cashing In on the Outdated iPhone, September 17. www.prweb.com/releases/sell/iphone/prweb1349134.htm.

Quine, W. v. O. 1936. Truth by Convention. In *Philosophical Essays for A. N. Whitehead*, ed. O. H. Lee. New York: Longmans.

———. 1960. Carnap and Logical Truth. *Synthese* 12, no. 4: 350–74.

Reuters. 2006. Spitzer Says AOL Woes Remain an Issue, July 10. www.ukfast.co.uk/internet-news/spitzer_says_aol_customer_woes_remain_an_issue.html.

Rogers, Everett M. 2003. *Diffusion of Innovations, Fifth Edition.* New York: Free Press.

Schatz, Amy. 2005. Crossing the Digital Television Divide. *Wall Street Journal,* July 6.

Seideman, Tony. 1993. Barcodes Sweep the World. *Invention & Technology Magazine* 8, no. 4 (Spring). www.americanheritage.com/articles/magazine/it/1993/4/1993_4_56.shtml.

Shor, Mikhael. 2006. "Brief Game Theory History," *Game Theory .net.* www.gametheory.net/dictionary/game_theory_history.html.

Swedberg, Richard. 2001. Sociology and Game Theory: Contemporary and Historical Perspectives. *Theory and Society* 30, no. 3 (June): 301–35.

van den Bulte, Christophe. 2000. New Product Diffusion Acceleration: Measurement and Analysis. *Marketing Science* 19, no. 4 (Autumn): 366–80.

Varian, Hal. 2001. High Technology Industries and Market Structure, with Commentaries by Erik Brynjolfsson and Kevin M. Murphy. http://people.ischool.berkeley.edu/~hal/Papers/structure/.

von Neumann, John, and Oskar Morgenstern. 1944. *Theory of Games and Economic Behavior.* Princeton, NJ: Princeton University Press.

Wingfield, Nick. 2006a. A New Reason to Buy a High-Definition TV. *Wall Street Journal*, January 4.

———. 2006b. Format Faceoff: Bringing the DVD War Home: Blu-ray Player and Titles Hit Stores, Taking On HD-DVD; Is It Time to Choose Sides? *Wall Street Journal*, June 20.

Index

D

E

F

G

H

I

S

T